Challenges in Ceramic Product Development, Manufacture, *and* Commercialization

Related titles published by the American Ceramic Society:

High-Temperature Ceramic-Matrix Composites II: Manufacturing and Materials Development (Ceramic Transactions Volume 58)
Edited by A.G. Evans and R. Naslain
© 1995, ISBN 0-944904-99-8

High-Temperature Ceramic-Matrix Composites I: Design, Durability, and Performance (Ceramic Transactions Volume 57)
Edited by A.G. Evans and R. Naslain
© 1995, ISBN 0-944904-98-X

Design for Manufacturability of Ceramic Components (Ceramic Transactions Volume 50)
Edited by Asish Ghosh, Basavaraj Hiremath, and John Halloran
© 1995, ISBN 0-944904-88-2

Manufacture of Ceramic Components (Ceramic Transactions Volume 49)
Edited by Basavaraj Hiremath, Allan Bruce, and Asish Ghosh
© 1995, ISBN 0-944904-87-4

For information on ordering titles published by the American Ceramic Society, or to request a publications catalog, please call 614-890-4700, or write to Customer Service Department, 735 Ceramic Place, Westerville, OH 43081.

Challenges in Ceramic Product Development, Manufacture, *and* Commercialization

▼▲▼

Edited by
Asish Ghosh
Philips Display Components
Ronald E. Barks
Ronald E. Barks & Associates
Richard P. Rusin
3M Corporation

▲▼▲

Ceramic Transactions
Volume 66

Published by
The American Ceramic Society
735 Ceramic Place
Westerville, Ohio 43081

Proceedings of the symposium Challenges in Ceramic Product Development, Manufacturing, and Commercialization, held at the 97th Annual Meeting of the American Ceramic Society, April 30–May 3, 1995, in Cincinnati.

For information on ordering titles published by the American Ceramic Society, or to request a publications catalog, please call 614-890-4700.

Printed in the United States of America.

1 2 3 4–00 99 98 97 96

ISSN 1042-1122
ISBN 1-57498-010-6

Contents

Commercialization of Bioceramics

1995 Corporate Award Winners

General Issues

Financial Issues in Commercialization

INNOVATIVE TECHNOLOGY TRANSFER INITIATIVES

Arthur Ballato and Richard Stern
US Army Research Laboratory
Fort Monmouth, NJ 07703-5601

ABSTRACT

The Physical Sciences Directorate (PSD) of the Army Research Laboratory (ARL), Fort Monmouth, NJ, has achieved significant success in the technology transfer/strategic alliance arena by blending technology transfer statutes, regulations and practices into its corporate culture. This paper describes the process and procedure PSD utilized in achieving that success, along with the lessons learned in introducing new and innovative methodologies, techniques, and approaches in transferring federal laboratory technology to the private sector.

INTRODUCTION

Physical Sciences Directorate is one of eleven directorates of the Army Research Laboratory, with headquarters in Adelphi, MD. The Directorate consists of 239 individuals, 177 of whom are scientists and engineers (S&Es), including 59 PhDs. PSD is the Army's focal point for research and development in the physical sciences, and generates more than 50 issued patents each year. It provides enabling technologies to solve critical barrier problems in photonics, opto-electronics, micro-electromechanics, smart materials, solid-state and nanoscience, electrochemistry, energy science, bioscience, high-frequency electronics, RF acoustics, manufacturing science, and electrophysical modeling, advancing the physical sciences technology base consistent with combat needs of the Army.

PSD has developed a well-focused, broad-based technology transfer program that creates both strategic and tactical alliances with business, industry, and academia. The Directorate is committed to an "open laboratory" policy where entrepreneurial S&Es from the private sector may engage in on-site cooperative efforts with PSD's S&Es, utilizing the Directorate's intellectual property, unique and sophisticated facilities and high level of expertise to develop or improve commercial products and processes of interest to both the private sector and the Army.

PSD has implemented more than 60 Cooperative Research and Development Agreements (CRDAs) and 18 Patent License Agreements (PLAs) since the Technology Transfer Act became law in 1986, with additional agreements in various stages of preparation. An income stream, established from of these agreements, is being used to expand the PSD technology transfer program effort, cover CRDA operating expenses, and reward contributing S&Es.

Our goal is to exploit the latest scientific advances to meet military needs while, at the same time, fostering the creation and improvement of commercial products and services within the civilian economy to foster both the economic and military success of the United States.

CHANGING THE CULTURE

PSD management recognized, early in the 1970s, that the preponderance of R&D resources and technology needed by the Army existed outside the organization. Management therefore determined that it was essential for PSD to identify and interact with the best available outside research resources; this could only be done successfully if the interaction was structured to be to the mutual benefit of all concerned. This approach required individual S&Es to develop an awareness of and familiarity with research and researchers throughout the world.

Towards that end, S&Es at PSD are required to complete a minimum of 80 hours of annual training in the areas of technology, technology transfer, intellectual property, and marketing. PSD provides a continual string of lecturers from business, industry and academia, who provide workshops, lectures and seminars in addition to the availability of regular college and university curricula.

As a result of this approach, informal cooperative efforts were initially developed between PSD and technical contractors, colleges and universities, and through technical working relationships borne out of personal interactions established at technology conferences, symposia, seminars, workshops, etc. The aim of these teaming efforts was to address specific technical barrier problems using a larger and a more diverse R&D work force that approached a critical mass. The advent of the Technology Transfer Act of 1986 and the related Executive Orders of 1987 provided, at that point in time, a mechanism for formalizing, expanding, and strengthening informal cooperative efforts that already existed at PSD. The first CRDAs established at the Directorate were carried out with organizations with which PSD had informal or contractual relationships. This approach minimized difficulties and obstacles involved in establishing the initial formal collaborative agreements. Implementation of each of these new CRDAs, however, required continual nurturing by the PSD Technology Transfer Office, otherwise they would have languished.

DEFINING TECHNOLOGY TRANSFER

One must first define what is meant by "technology transfer" before describing the PSD approach to the implementation and management of a technology transfer program. This is sometimes best accomplished by explaining what technology transfer is not.

Effective technology transfer is not merely the process of making known the availability of technology. It is not "throwing technology over the transom," for others to pick up and possibly utilize without further explanation or discussion. Little is accomplished by this

hands-off approach, and disappointment in the process and its accomplishments will surely follow.

Some consider the transfer of federal funding to the private sector for the development of dual-use technology a form of technology transfer. Although useful technology is developed through contractual efforts, this should not be considered transfer of existing federal laboratory technology for the benefit of both the laboratory and the private sector.

Establishing a CRDA for its own sake without specific, focused objectives, no matter what resources are made available for that collaboration, cannot be considered a potentially productive technology transfer effort. Without an objective and statement-of-work beneficial to all parties and to which all parties are committed, disappointment can be expected.

A fully functional technology transfer program is one where S&Es are motivated to transfer/transition federal laboratory technology (in-house technology, methodologies, expertise, capabilities, facilities use) to the private sector (business, industry and academia) by means of a CRDA or PLA in a way that benefits all participating parties. The process should be reciprocal, with both federal and non-federal partners contributing to the effort. A win-win situation should be a driving objective.

TECHNOLOGY TRANSFER ADVANCEMENT

There are a number of positive steps that every federal laboratory can take to promote technology transfer. First, it is important that a laboratory, intending to develop an effective technology transfer program, have leadership which champions technology transfer. A laboratory having management unsympathetic to the cause is one where the technology transfer program will neither reach its potential, nor satisfy any of those involved. Further, the lab must foster a strategic view of the technology transfer process, seeing technology

transfer as a tool used by all elements of the laboratory to assist the lab in meeting its mission goals. An environment supportive of technology transfer needs to be created. Led by example, S&Es trained in the generation and use of CRDAs and PLAs by a technology transfer manager and a user-friendly legal staff, will understand the value of intellectual property, and appreciate the value of cooperative efforts. By encouraging S&Es to attend and network at technical meetings, conferences, workshops and seminars and to generate working relationships with their peers outside of their laboratory, even marketing soon becomes a distributed activity. The use of database searches, publishing and presenting technical papers and S&E exchanges all contribute to the open laboratory concept.

Certain ingredients are necessary for operating a successful technology transfer program. Among these are: motivation; flexibility; trustworthiness; unique technical expertise; infrastructure equipment and easy access thereto; identification of technology ripe for commercialization; realistic expectations; responsibilities clearly delineated; fast response time; local negotiation authority; simple procedures; reasonable CRDA and PLA terms; and user-friendly attorneys.

COLLABORATIVE RESEARCH AND DEVELOPMENT AGREEMENTS

PSD has now established nearly all of its technology transfer and collaborative efforts under formalized CRDAs and PLAs. These business-like contracts are useful because they define the technology transfer process including allocation of intellectual property rights, termination procedures, dispute resolution, publication rights, and limit they federal liability through indemnification.

Successful technology transfer efforts usually exhibit the following characteristics: participants have a strong vested interest in the outcome; all participants benefit in fair and equal portions; all parties are involved early-on; technology user/customer is involved in the

early stages, with simultaneous development of technology and product; and customer-oriented product is actively marketed.

Conversely, barriers to technology transfer include; perceived lack of fairness, suspicion or lack of trust, NIH syndrome/arrogance/pride, legalistic mindset, and short-term focus. Each can hinder successful technology transfer.

The PSD Technology Transfer Office processes, on the average, one new CRDA a week. Although the content of these CRDAs is very similar, each CRDA has its own personality. Some CRDAs involve large corporations, such as Martin Marietta or Texas Instruments, while others establish partnerships with small start-up businesses. A CRDA may be a supplement to an ongoing contract, allowing the Army to assist in the advancement of the R&D effort, or may be initiated in order to aid a company that needs help in commercialization under an existing PSD PLA. Some CRDA participants provide new and unique electronic materials to PSD in exchange for test and evaluation of those materials by PSD technologies utilizing sophisticated analysis equipment, facilities and techniques available at PSD. Some CRDAs represent a "spin-on" to the Army, where a small business wants and needs technical assistance in the development of a new patentable idea that may be useful to the Army as well as the private sector. Several of our CRDAs with academia (Princeton, Rutgers, Stevens and NJIT) involve placement of graduate students, post-doctoral resident associates and professors into the PSD laboratory where they have the advantage of working together with PSD S&Es using state-of-the-art facilities and equipment not readily available in academia. PSD has the advantage of enhancing its work force through the use of these individuals, thereby creating a win-win situation. CRDAs have also been used to formalize working relationships with industrial partners in the creation of collaborative team efforts in carrying out ARPA TRPs (Technology Re-investment Programs) or other federal agency contract awards.

When establishing a CRDA, PSD particularly looks for specific elements, including a mission related project, a win-win situation, a leveraging opportunity, income, dual-use applicability, and the potential for a follow-on PLA.

A PSD CRDA consists of approximately 12 pages, and can be created in as little time as a few days, with Army formal approval received in fewer than 30 days. In the absence of federal acquisition regulations governing CRDAs, the CRDA contract terms anticipate conflict situations, for example, defusing potential problems involving data rights and intellectual property. The CRDA is one of the most effective and perfected tools in the PSD technology transfer repertoire.

PATENT LICENSING

We find that patent licensing is a very efficient means of transferring technology, and we have granted licenses of PSD technology to both small and large companies. By entering the patent licensing market, we learned a number of valuable lessons.

The first and probably most important consideration in licensing is the breadth of the patent claims. The quality of patent claims is, in fact, that for which the licensee is paying. Claims that are not well crafted will not prevent others from "patenting around" your claims. Patents are too often rendered useless for licensing because of their claims are too narrow. This situation arose from the old mind-set that considered patenting from a purely defensive standpoint: it was only used to prevent others from charging the government royalties for its own research. Sometimes, however, an older patent with typically narrow defensive claims can be rehabilitated by applying for new and related patents that expand around (picket fence) the existing claims and uses of the existing invention/patent. This is part of patent portfolio management.

Second, federal laboratories should be sensitive to the commercialization requirement (Code of Federal Regulations 37, Part

404) of any patent license they issue. This regulation requires that the licensee be capable of, and intend, to commercialize broadly the government patent in a reasonable time frame. This is a condition which is not difficult for a large corporation to meet. Often, however, when dealing with a small business, where this may be a problem, a federal laboratory should lean toward issuing non-exclusive licenses, at least until the licensee can show, either by internal growth or strategic partnerships, the capability of supplying the market place with the licensed technology. Further, milestones need to be a part of the patent license in order to insure that the licensee is bringing the technology to market without delay. Failure to meet milestones can be a clear indication that effective commercialization will not be achieved and the license can either be terminated or converted to a less exclusive mode so that others may have the opportunity to license and advance commercialization. Options for greater exclusivity incorporated into non-exclusive licenses can be a way to allow, at a later date, a small business licensee to gain exclusivity through exercise of the option when the small business is in a better position to prove its ability to commercialize the technology.

The PSD Technology Transfer Office usually requires an up-front fee when licensing its technology to business and industry. This fee is charged to cover partially the cost of applying for the patent and the related maintenance fees and also as a measure of good faith of the licensor. Up-front licensing fees charged to small business are usually paid in increments distributed over a period of years. Royalty rates vary with the technology field and can sometimes be partially written off against the initial up-front licensing fee. Knowing that the bottom line is the timely commercialization of the technology, we utilize flexibility in licensing terms to bring PSD technology to the market as quickly as possible.

Each of the patent license issues raised here involve "negotiations." Negotiation is a talent which must be developed through education, good legal advice and experience. This seems to be one of the most difficult aspects of technology transfer with which federal technology

transfer offices have to deal. Fortunately, federal laboratories often have more maneuver room in negotiating than private sector firms because our first priority is technology commercialization.

TECHNOLOGY MARKETING

There are many means by which we can and do market our technology. We showcase our technology at technical expos, advertise in trade journals, utilize technology brokers, distribute booklets describing PSD patents available for licensing, provide inputs to major databases which are broadly available to the public, publish in the Federal Register, examine citations to our patents that appear in other newly issued patents to expose potential areas of infringement, publish and present technical papers at conferences and symposiums, sponsor workshops and seminars and have our technologists network with their peers in the private sector. By far, our major marketing successes have been achieved through contacts made by our technologists networking with their peers in business, industry and academia. Most of our S&Es now operate in an entrepreneurial-like mode in dealing with potential technology users in the private sector. The results indicate that our S&Es are highly motivated to bring in new technology partners/users; act as consultants to the private sector; and deliver unique prototype designs, components and devices to business and industry. Our technologists, understanding the benefits of technology marketing, have changed the way PSD does business.

Examples of PSD CRDA's pertinent to the area of ceramics include:

- **"Laser Ablation of Ferroelectric and High-Temperature Superconducting Thin Films," Rutgers University, New Brunswick, NJ**
- **"Development of Smart Ceramic Materials," Rutgers University, New Brunswick, NJ**
- **"Development of Hermetic Coatings for Optical Waveguides," Rutgers University, New Brunswick, NJ**

- **"New Piezoelectric Materials with Application to Frequency Control," RF Monolithics, Dallas, TX**
- **"Development of a Permanent Magnet System for a Microwave Tube," Martin Marietta, Rancho Bernardo, CA**

PSD has spent several years developing an effective, efficient and productive technology transfer program, so that today, the major effort of the PSD Technology Transfer Office is concentrated on managing the program rather than its development. Areas such as marketing, licensing, negotiation, program tracking, exploring innovative approaches to technology commercialization and managing our intellectual property portfolio have now become the main tasks of the Technology Transfer Office at PSD.

THE SBIR PROGRAM AND TECH TRANSFER

PST Technology Transfer Office has developed a new and innovative method of promoting the commercialization of its technology. We are utilizing the Small Business Innovation Research (SBIR) Program to address the development, licensing and commercialization of PSD patents. Proposals are solicited which are aimed at productizing specific PSD patents which have already been prototyped and demonstrated by Army technologists/inventors. These prototype designs still require commercially oriented development and optimization, with consideration given to manufacturing and transitioning into actual utilization for both military and private sector applications. SBIR contractors who successfully productize the subject patents are licensed by the Army to produce and market the technology. This program appears to yield significant advantages. First, the small business is given funding to commercialize a relatively low risk invention. This productization venture could in turn become the mainstay and contribute to the healthy growth and success of that particular company. Second, those inventions chosen to be commercialized under the SBIR program are hand-picked, with strong consideration given to the need for that particular technology in both the military and private sector. Third, this methodology

provides for the development of "niche" technologies where large corporations either have no capability or desire to initiate development efforts and where small businesses have no funding base to pursue the technology development on their own. Lastly, using this approach, Army technology is fully developed, demonstrated and effectively utilized for commercial applications, with patent licensing fees and royalties being paid to the inventors and the Army laboratory where the invention was conceived.

This new commercialization program effort is now in progress under two SBIR contracts and is advancing well. Even before completion of the SBIR contracts, the small business contractors are already producing results, and expect to start delivering on orders from the private sector in the near future.

PERSISTENCE

One of the most important facets of a well-oiled technology transfer program is follow-up. Follow-up on personal contacts made at expos, conferences, seminars, etc., leads to new partnering arrangements under CRDAs and PLAs. Tracking milestones, royalty payments, commercialization plans, etc., with respect to existing patent licenses insures that laboratory technology is well on the road to commercialization. Follow-up on progress and payments (if required) under existing CRDAs, and follow-up on investigative measures taken to identify possible infringements on Army patents are equally important. Continuous follow-up cannot be overemphasized.

CONCLUSION

Once a federal laboratory technology transfer program has been successfully developed and in full operation, one cannot assume that technology and money will flow automatically. Marketing efforts must expand apace and technologists must be continually educated in the value of their technologies and their intellectual property. S&Es must be encouraged to become marketeers, while attorneys must be made

to be more user-friendly. All participants: legal, technology-transfer, and marketing must learn to be better communicators, in order to nurture more effectively and harvest more fully the fruits of the strategic alliance that has become our new culture.

REFERENCES

[01] Stevenson-Wydler Technology Innovation Act of 1980 (PL 96-480)

[02] Bayh-Dole Act of 1980 (PL 96-517)

[03] Small Business Innovation Development Act of 1982 (PL 97-219)

[04] Cooperative Research Act of 1984 (PL 98-462)

[05] Trademark Clarification Act of 1984 (PL 98-620)

[06] Japanese Technical Literature Act of 1986 (PL 99-382)

[07] Federal Technology Transfer Act of 1986 (PL 99-502)

[08] Malcom Baldrige National Quality Improvement Act of 1987 (PL 100-107)

[09] Exec. Orders 12591 & 12618 (1987): Facilitating Access to Science & Tech.

[10] Omnibus Trade and Competitiveness Act of 1988 (PL 100-418)

[11] NIST Authorization Act for 1989 (PL 100-519)

[12] Water Resources Development Act of 1988 (PL 100-676)

[13] National Competitiveness Technology Transfer Act of 1989 (PL 101-189)

[14] Defense Authorization Act for FY 1991 (PL 101-510)

[15] Intermodal Surface Transportation Efficiency Act of 1991 (PL 102-240)

[16] American Technology Preeminence Act of 1991 (PL 102-245)

[17] Small Business Technology Transfer Act of 1992 (PL 102-564)

[18] National Department of Defense Authorization Act for 1993 (PL 102-25)

[19] National Defense Authorization Act for FY 1993 (PL 102-484)

[20] National Department of Defense Authorization Act for 1994 (PL 103-160)

[21] Army Regulation 70-57, "RD&A Military-Civilian Technology Transfer," 7/91

[22] R. Stern and T. Wittig, "Technology Transfer: Lessons Learned - Preparation for the Future," Proceedings - Technology Transfer Society Annual Conference, Washington, DC, July 1995, in press.

ACCESSING PUBLIC LABORATORIES TO IMPROVE COMPANY TECHNOLOGY:

presented by Richard B. Cass, President of Advanced Cerametrics, Inc., Lambertville, N.J. May 2, 1995

BACKGROUND OF ACI

My company, Advanced Cerametrics, (ACI) produces a range of advanced ceramic materials for many industries. ACI is the result of a 1990 merger of 45 year old Lambertville Ceramic and four year old HiTc Superconco.

During the past seven years ACI has been active in several areas of technology development and has participated in four CRADA's with Doe Labs and has been the recipient of several SBIR Phase I and Phase II Awards, STTR grants and other government sponsored R & D efforts.

ACI joined the newly formed Association of American Ceramic Component Manufacturers (AACCM) in 1992. This group is made up of U.S. ceramic manufacturing companies that have a common goal "to expand the markets for manufactured ceramic components by enhancing the process and product quality; and by increasing public and industry education and awareness of ceramic applications." this group currently has 35 member companies with increasing enrollment which is open for any qualifying company. Membership in AACCM has benefitted each company by increasing inter-company purchases, inquiry referrals, and exchange of information on markets, competition, supplies, and employee benefits including wages.

HISTORY OF CRADA

In 1993 AACCM was approached by Sandia National Lab with the idea to participate in a cooperative research and development agreement (CRADA) to adapt powder compaction modelling techniques to contain spent nuclear fuel developed by the D.O.E. to use in the ceramic industry. After almost two years of negotiations this CRADA is now in operation. There are six member companies involved in the CRADA and Los Alamos and Sandia are conducting the basic research. The CRADA is expected to take five years to complete with the final result being a computer modelling program for predictive analysis of pressing, forming, binder removal and sintering of ceramic parts. The concept will lower the cost of production of ceramic artifacts and make the member companies more competitive in the world market.

This industry/CRADA group is called the AACCMCI or ACME for short and is in a position to grant licenses to American companies through out the development process and after it is complete. I was asked to head up the new group because of our company's experience in dealing with government labs.

We are told that this CRADA is the first time that more than one competing government lab has cooperated with an industry consortium to commercialize technology.

The difficulties encountered during the above negotiations are the subject of this presentation.

BACKGROUND TO THE NEGOTIATION OF THE CRADA

BENEFITS

- AACCM was approached by Sandia to attempt to commercialize technology which had been developed at taxpayer's expense, but had little or no commercial value.
- The proposed idea of making a computer model to predict pressing problems before a tool was even designed has great appeal for our industry for cost savings, reducing material loss and the unfortunate situation, which we have all experienced, of rebuilding a tool several times until it makes parts without cracks or other defects.
- The idea of cooperating with a large government lab had immediate appeal for several AACCM member companies because of the availability of resources outside of the proposed CRADA and frankly the romance of having such an arrangement on the company resume.
- The internal work that the member companies are obligated to perform in order to develop the technology will provide introspective analysis of each company's internal systems and operations with the net result of improving them.
- It is possible that an added benefit to the member companies will include downstream royalties from licensing of the jointly developed intellectual property.

For further information regarding membership in AACCM, AACCMCI or licensing in the ceramic compaction modelling CRADA please contact Janie Deibler at the American Ceramic Society at 614-794-5861.

DIFFICULTIES IN THE CRADA NEGOTIATION PROCESS:

- Because each lab is fighting for a diminishing pool of research dollars only the highest profile work and/ or corporate partner are given priority. Congress is more likely to authorize funds to a project that it is familiar with.

- Cessation of the Technology Reinvestment Project (TRP) funding has put a hold on many programs which would have been funded three years ago.

- "Funds in CRADA" - which means the corporate partner pays for all of the work, both at the lab and in its facility - are still accessible, but funding a government lab is an expensive proposition.

- Intellectual property issues are thorny and can often delay or prevent the signing of an agreement. The government reserves fairly liberal rights regarding technology developed jointly.

- The administrative bureaucracy at the labs is enormous and can delay ratification of an agreement for months, often pushing an agreement into a new fiscal year where funding may no longer be available for the CRADA.
- All of these concerns are driven by money and it is always possible that at the last minute in your CRADA's negotiation someone high up in the lab's administration may divert the required funds for a more visible program which will bring federal funds to the lab the CRADA negotiations with Sandia and Los Alamos that AACCMCI went through had all of the above elements, plus a change in administration which lost the focus of our project and the only way we got it back on track was to threaten the labs with law suits and contact with our congressmen.

Since our CRADA is not a "funds in" type of agreement we have had to depend on the labs to find their own money for the program. This has truly been a nightmare.

As an over view, I would like to take a brief look at the status of all government funded CRADAs and which department is the most likely to reach agreement with a business or consortium.

Over the past seven years the government labs' goals and approach to industry have changed from

- "If you want us to do research for you, either pay for it or do it the way we want to"

- To "dual use technologies" which said to industry 'look guys we have this wonderful technology here that we are doing for our own programs, how about buying a license for your uses and helping us pay for our programs since Washington probably won't any more"

- To today's situation where the labs have to try and sell what they have developed for the government to industry or risk the closure of the individual research programs and potential loss of employment by the researchers.

The much heralded Technology Reinvestment Project (TRP) which was initiated to fund and promote the current cooperation with the labs has been described by the house appropriations committee as a failure with the claim that "despite significant levels of (TRP) funding ($1.2 billion since 1993) the defense department has yet to identify any military benefit from pursuing this program"

CRADAs are an excellent vehicle to access the technologies developed over the past fifteen years. They make the tech transfer process fairly simple, but unless the money is in place, forming a CRADA alliance will still be unlikely because each of the labs and its programs are fighting for a diminishing supply of dollars.

This is an unfortunate situation because there are certain to be thousands of developments and mature technologies that could be of benefit to American industry that will be held behind closed doors and eventually forgotten because the budget ax has reduced the required lab participation to a level that will only permit the dissemination of those technologies that have the most immediate hope for commercial return on investment.

CONCLUSION

From personal experience I recommend that unless your company or group is willing to front a significant portion of the cost of the technology transfer or technology development from the labs you should not bother. As much as the labs or the lab personnel may want to work with you they will probably be unable because of funding limitations.

SYNCHRONOUS FLOW MANAGEMENT - HOW TO WIN NOT MERELY SURVIVE

S. O. Roberts, SFM Sunapee, Wooster, OH; R. B. Vollum
R. B. Vollum & Associates, Huntingdon Valley, PA

In the war for global economic survival, the luxury of yesterday's answers is not acceptable. We can not afford to play catch-up, but must leapfrog ahead with new management ideas and technology. To remain viable and competitive, a company must adopt a rigorous and structured process of ongoing improvement. Synchronous Flow Management techniques coupled with the Theory of Constraints philosophy and thought structure provide such a process. They force the critically important issues into focus for management attention, and insure that resources are employed where they will have the most positive economic impact.

The business and operational philosophies of Synchronous Flow Management, formulated in the 1980's by Eliyahu M. Goldratt and developed further by Robert B. Vollum, have become an integral part of manufacturing companies across the United States.

Synchronous Flow Management is any systematic way that moves "the process" and product flow quickly and smoothly through the various resources of a business in concert with market demands. This would appear to be an obvious and common sense statement; however, a company has a major opportunity to significantly improve its business performance with some very simple tools. This treatise is based upon companies who have a Vision, Mission Statement, Strategy, and Business Plan in order to maximize the benefits of embracing Synchronous Flow Management. Synchronous Flow Management also requires the support of upper management and the education of all employees, with a clear plan to address the inherent obstacles of change and implement the following procedural steps:

1. Universal, individual ownership of ideas
2. Group consensus that Synchronization will work for you
3. Where and how much will Synchronization improve operations?
4. What needs to be changed and what will be the results?

5. How do you do it?
   - Implementation Plan
   - Education
   - Guidance
6. Whose job is it?

What is the Goal of any company? Is it better customer service, larger market share, lower cost, improved quality, survival? Though all of these are important to any business, the Goal is to make money in the present as well as in the future. If the Goal is to make money, what are the bottom line measurements:

NET PROFIT - absolute
RETURN ON INVESTMENT - relative
CASH FLOW - survival

The bridge to these global measurements is the action of management.

Synchronous Flow Management requires that the cost concepts be replaced with global operational measures and a change to the paradigm in which each manufacturing organization currently operates. These measure are defined as follows:

Throughput - The rate at which the system generates money through sales. You can get creative and drive the world to you, given the limit of existing resources.

Inventory - All the money the system invests in purchasing things the system intends to sell. There is a limit, but the reduction of inventories is strategic to Throughput.

Operating Expense - All the money the system spends in turning Inventory into Throughput.

The intuitive goal of every manufacturing organization is to increase Throughput, decrease Inventory, and decrease Operating Expense simultaneously. The strategic role for reduced inventory concerns the issues of product (quality, engineering), revenue (increased Throughput, lower investment per unit), and responsiveness (on time delivery performance, shorter lead times); they all represent the COMPETITIVE EDGE.

All organizations are controlled by constraints. Constraints are anything that prevent a system from achieving a higher level of performance relative to the goal. Constraints can be good; constraints can be bad; but constraints can be managed to our economic advantage. Constraints can be Physical (bottlenecks, overloaded resources, marketplace, vendors), Managerial (policy, measurement systems and incentives, perceived attitudes), and Behavioral (work habits, attitudes, labor/management relationships, predominant work ethic). In order to manage these constraints there are 5 procedural steps:

1. Identify the system's constraints.
2. Decide how to exploit the system's constraints.
3. Subordinate everything else to the decisions made in step 2.
4. Elevate the system's constraints.
5. If the level of the constraint is elevated to the point that the constraint is broken, go back to step 1 and start the process again.

The Throughput world of Synchronous Flow Management uses measurements that are extremely powerful:

Throughput = Sales - Variable expense (raw material)
Net Profit = Throughput - Operating Expense
Return on Investment = (Throughput - Operating Expense)/Inventory
Inventory Turns = Throughput/Raw Material
Productivity = Throughput/Operating Expense

The traditional cost world places emphasis on the allocation of cost tied to direct labor and product cost. These cost procedures constrain Throughput, increase Inventory, and balance resource capacity. In the traditional cost world operating expense is the highest priority, with Throughput and Inventory a distant second and third. The Throughput world places emphasis on Throughput, Inventory is strategic to Throughput, and Operating Expense must NOT be managed to constrain the Throughput of the business.

Synchronous Flow Management is governed by 9 Rules:

1. BALANCE THE FLOW, NOT THE CAPACITY
   The balancing of any manufacturing resources (equipment, people) constrains any business due to statistical fluctuations and dependent events.
2. THE UTILIZATION OF A NON-BOTTLENECK IS NOT DETERMINED BY ITS OWN POTENTIAL, BUT BY SOME OTHER CONSTRAINT IN THE SYSTEM
   All manufacturing organizations have a bottleneck (capacity constrained resource), or point of control (most constrained resource). The utilization of a non-bottleneck resource is determined by the constrained resource and its utilization.

3. ACTIVATION AND UTILIZATION OF A RESOURCE ARE NOT SYNONYMOUS
   A resource may be 100% utilized, but may be activated at a rate far below its capacity.
4. AN HOUR LOST BY A BOTTLENECK IS AN HOUR LOST FOR THE TOTAL SYSTEM
   A bottleneck controls Throughput and any loss of time is a corresponding loss of Throughput.
5. AN HOUR SAVED AT A NON-BOTTLENECK IS JUST A THROUGHPUT MIRAGE
   An hour saved at a non-bottleneck may or may not have value, but will not increase Throughput.
6. BOTTLENECKS GOVERN BOTH THROUGHPUT AND INVENTORIES
   The proper identification and management of bottlenecks will maximize Throughput and minimize Inventory.
7. THE TRANSFER BATCH MAY NOT, AND MANY TIMES SHOULD NOT BE EQUAL TO THE PROCESS BATCH
   It is not possible to maximize Throughput, reduce inventories, and reduce process time if the operating philosophy calls for large process batches, and no flexibility to properly flow material using smaller transfer batches.
8. THE PROCESS BATCH SHOULD BE VARIABLE AND NOT FIXED; VARIABLE IN TIME AND ALONG ITS ROUTE
   Process flexibility will increase Throughput, decrease inventories, and reduce process/lead times.
9. SCHEDULES SHOULD BE ESTABLISHED BY LOOKING AT ALL THE CONSTRAINTS; LEAD TIMES ARE THE RESULT OF THE SCHEDULE AND CAN NOT BE PREDETERMINED
   Predetermined lead times are a self fulfilling prophecy; and on time delivery performance will not be acceptable. Competitiveness issues suffer and future profitability is compromised.

In summary applying the principles of Synchronous Flow Management is relatively straight forward. The results can be direct and fast; Throughput, Net Profit, Return on Investment, Cash Flow, On Time Delivery Performance, and Quality can dramatically improve while Inventory, Operating Expense, and Lead Time can be decreased. The key to Synchronous Flow Management is constraint identification and management: A NEVER ENDING PROCESS OF CONTINUOUS IMPROVEMENT.

Product Development and Manufacturing

AN INTEGRATED APPROACH TO PRODUCT DEVELOPMENT AND MANUFACTURING

Michael J. Readey
Glass and Electronic Ceramics
MS 0333, Department 1845
Sandia National Laboratories
Albuquerque, NM 87185-0333

ABSTRACT

A new approach to product development is described that integrates various unit operations into a unified "knowledge-base". This knowledge-base is easily accessible to all members of the design team due to the advent of high performance and networking capabilities of today's desktop computers. This permits rapid optimization of the product's material, shape, and manufacturing processes that satisfy the customer's performance requirements while maximizing economic return for the manufacturer.

INTRODUCTION

The markets for technical ceramic products have grown steadily during the past several decades, due in part to the emergence of higher performance materials such as zirconia and silicon nitride, but also because of lower cost components brought about by more efficient manufacturing. Much of this growth has been fueled by an insatiable quest for high quality, primarily driven by significant international competition. But high quality is no longer a distinguishing factor in the marketplace. For instance, programs initiated by most technical ceramic manufacturers have led to large improvements in quality, which in turn have resulted in considerable advances in component performance. At the same time, the level of quality and performance have become more uniform from one vendor to another. Hence a vendors competitive edge by claims of higher quality has largely disappeared. High quality is now simply a condition for doing business in the marketplace.

Another interesting trend is the customization of what was once considered high volume commodity products. Customers now desire products tailored to their specific applications. They may be purchasing lots of the same overall volume, but now of these lots will consist of smaller sub-groups of components having different features. Thus while ceramic manufacturers have anticipated growing markets that would allow them to transition from a traditional job-shop to more continuous flow operations, customers are expecting lower costs from the economies of scale of larger markets, but greater levels of flexibility available from a job-shop environment. Moreover, customers desire this greater flexibility and lower cost together with reduced delivery times. This new trend suggests that, product quality and performance being equal, markets in the future will go to those manufacturers who deliver customized products in record time and at lower cost. Thus a new set of tools are required to develop flexible manufacturing strategies to rapidly develop new products while simultaneously reducing cost.

THE MATERIALS SCIENCE PARADIGM - IS IT ENOUGH?

At some time in their education, most materials engineers are exposed to the classic Materials Science Paradigm, schematically illustrated in Figure 1. This paradigm relates product performance to three primary factors: (i) processing conditions (in the case of ceramics, this refers to packing density, sintering time and temperature, etc.); (ii) structure (crystal structure, electronic structure, microstructure, etc.); and (iii) properties (mechanical, electrical, chemical, etc.). It is now widely recognized that much of the advances in the performance of many ceramic components come from a greater understanding of the these three areas, and how these areas are interrelated.

However, performance advantages relate only to the technical success of a product. Whether a product achieves economic success is a different issue. For example, a ceramic material may have the requisite performance characteristics when tested under laboratory conditions, but may in fact fail in the marketplace because it may not be cost-competitive with other engineering materials; it may be difficult to manufacture into the desired geometry; it may have a deleterious environmental impact; or it may simply not achieve the desired level of reliability over the products lifetime. These issues are generally not addressed by the materials science paradigm, yet are the key features that lead to product realization. This indicates that product development strategies must go beyond this oversimplified picture that focuses on strictly material issues, and use models that incorporate the more non-technical factors that together determine a product's ultimate success in the marketplace.

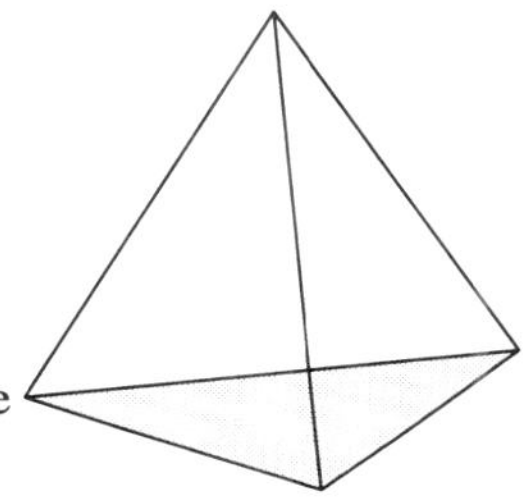

Figure 1. The classic tetrahedron illustrating the Materials Science Paradigm.

PRODUCT DEVELOPMENT MODELS

In order to understand the key features that lead to rapid product development, it is useful to first examine some of the models that exist today.

(1) The Linear Model

Historically, product design and development have followed a linear model, schematically illustrated in Figure 2. In this scenario, the Customers requirements are first received or interpreted by the Sales Engineer. If the requirements demand a new product design, sales relays the requirements to the Design Engineer. Depending on the similarity of the new component relative to existing products, the new design may then travel to the Manufacturing Engineer for advice on the appropriate manufacturing approach. A Process Engineer may be called in to assess modifications of existing processes to manufacture the new component. Finally, if there are material concerns, then the Materials Engineer becomes involved. Eventually answers work their way back up the chain, but often the process requires several iterations before all are satisfied. If all goes well, the new component has the same requirements initially requested by the customer. Unfortunately, communication is not always transferred with complete accuracy, and often times the customer ends up with something quite different than expected. Even if all does go well, the process can takes months if not years to complete, depending on the complexity of the component desired.

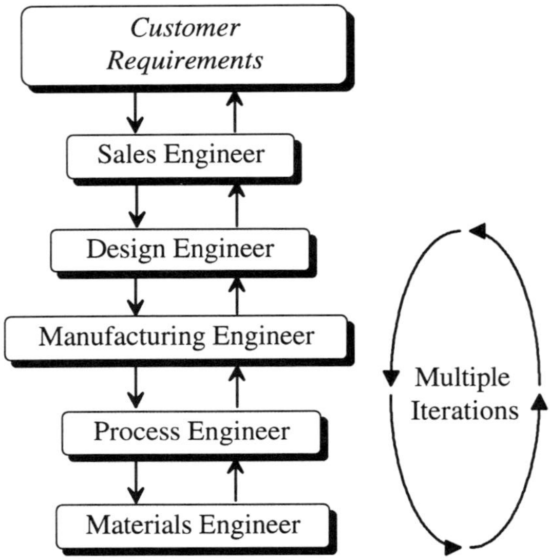

Figure 2. The traditional, linear model for product development.

(2) Concurrent Engineering Model

A significant improvement to the linear model evolved throughout the 1970's and 1980's, and is often referred to as the Concurrent Engineering Model, schematically shown in Figure 3. In this model, it is recognized that coordinating many efforts such as process engineering and materials engineering in a parallel fashion, or concurrently, significantly decreased the time required to complete a project. Thus, while the primary point of contact with the customer is still the Sales Engineer, the customers requirements are relayed to the entire engineering team at the same time, allowing for immediate discussion on the best possible approach towards achieving the customers requirements. This provides for a more direct path towards the final product, but also allows for "on-the-fly" design changes to be made without going through the entire development loop. However, despite the obvious advantages of this model of product development, it still has several disadvantages. For example, it is not uncommon for the product development "team" to disband once the project is completed. Thus much of the expertise learned during the development effort for that product is not retained, other than through the more typical and inefficient path of project reports. This can be a particular problem when team members retire or move to a different company altogether. Moreover, the process can still take weeks or months before the Sales Engineer can ultimately supply the customer with the desired product.

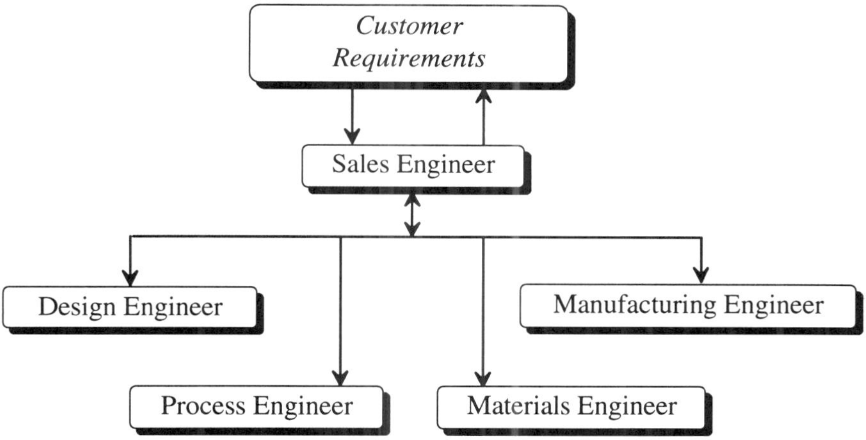

Figure 3. Concurrent engineering model for product development.

(3) Integrated Model of Product Development

While the concurrent engineering model is clearly superior to the linear product development model, it still lacks some key features. Hypothetically, the ideal model would incorporate the interdisciplinary nature of the concurrent engineering approach, yet be more quantitative in nature, as opposed to a purely heuristic-based approach. A more quantitative model would permit advanced computational analysis to be utilized in a "virtual" design and manufacturing environment, thereby optimizing the products characteristics and manufacturability before a single prototype is made. This would not only reduce the development time, but also greatly reduce costs associated with the trial and error method of product optimization. Finally, a quantitative model would establish a scientific and engineering database that could in principle link the areas of process technology, materials properties, manufacturing technology, and even economic and environmental assessment of a given design and manufacturing strategy. Such a model is schematically illustrated in Figure 4, and emphasizes that the various development efforts are integrated or linked together. In this case, the customer requirements form the focal point at the center, easily accessible by all team members. The project goals are defined at the initial stage of development by the interdisciplinary team, and are always cross-referenced throughout the effort in order to ensure that the project does not deviate from its initial objectives. Furthermore, the integrated nature of this model ensures that progress in one discipline serves as input to other disciplines, thus continually updating and advancing the development process.

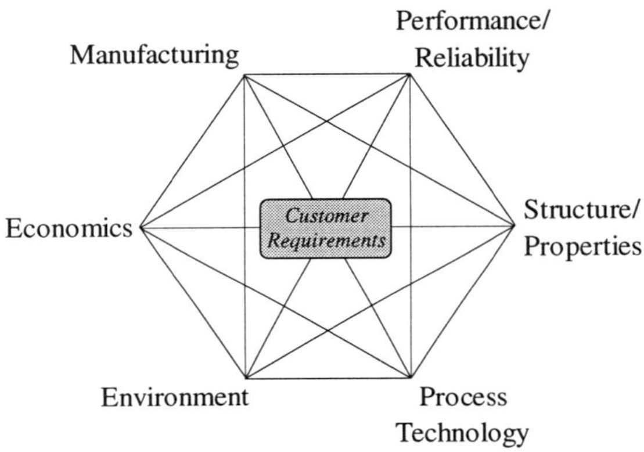

Figure 5. Integrated model for product development.

If the results within any one discipline were quantitative, for example models describing material properties as a function of processing condition (*i.e.*, sintered density as a function of time and temperature), then such information could be quickly accessed by other members of the team through the use of current computer networking systems. Each discipline could then incorporate the mathematical models of other disciplines into its own models, and so forth. The end result is that information once specific to a particular development task now becomes integrated into all other tasks, *i.e.*, the task-specific information databases become a single, unified project-level knowledge-base, as shown schematically in Figure 6.

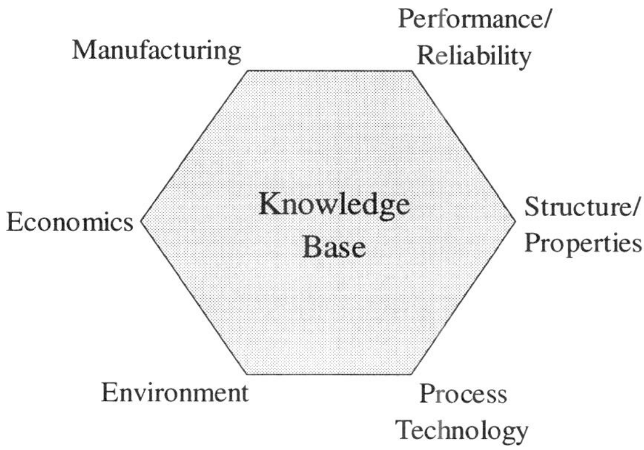

Figure 6. Concept of a knowledge base that integrates design disciplines.

A key feature of the knowledge-base is that it is retained long after the project has been completed; the quantitative nature of the model allows it to be stored for future access. For example, given a new product or design, the design and development team (or even the Sales Engineer alone) can easily conduct a comparative search of all similar products and processes, thereby circumventing the time and costs of repeating any one project from scratch. Using the existing knowledge-base as the starting point, the design and development team runs additional experiments and simulations relevant to the new product. This in turn leads to enhanced expertise, which can be schematically represented by the knowledge-base expanding outwards, as shown in Figure 7.

The expanding knowledge-base concept also provides a useful framework for considering the natural progression of manufacturing relatively simple components to those that are very complex. For instance, simple components, which are referred to as Level 1 components in Figure 7, do not require extensive knowledge to create. However, manufacturing Level 2 components, representing greater complexity, requires building on the expertise gained by fabricating Level 1 components. In other words, these more complex components could *not* have been produced without first producing the simple ones, or at least without a great deal of time and effort. Thus each level of complexity requires a specific knowledge-base; complex components generally require considerable expertise to fabricate cost effectively. The key point here is that building these complex components requires attaining the necessary knowledge-base by first fabricating less complex components.

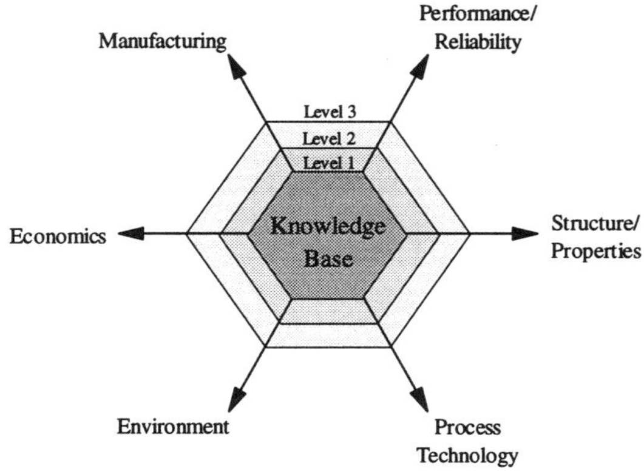

Figure 7. Expanding knowledge-base allows for the manufacturing of increasingly complex components.

A classic example of this last point concerns the advanced ceramic industry in the early 1980's, particularly in the structural ceramics area. At this time, the market for automotive components was expected to increase significantly due to the emergence of high strength, highly reliable ceramics such as zirconia and silicon nitride. Much of the applications involved high temperature components such as turbocharger rotors, extremely complex components having thick wall shafts and thin-walled, curved vanes. Several ceramic companies both in the U.S. and abroad and specifically Japan initiated development efforts to capture this rapidly expanding market. However, the approach to the development efforts in the U.S. and Japan were much different. While several U.S. firms attempted to manufacture the complex rotor directly, many Japanese firms developed manufacturing expertise by first fabricating fairly simple shapes, cylinders, balls, etc., while also working on the rotor itself. At the same time, high volume manufacturing techniques were established by selling fairly pedestrian consumer items such as ceramic scissors and sushi knives, products that are relatively simple in shape and design. Many in the ceramic industry considered such applications as frivolous. Yet these simple consumer items introduced the public to high performance ceramics, helping to foster greater confidence in designing with these materials for other advanced applications. But perhaps equally as important, these consumer items allowed manufacturers to optimize material properties and establish the processing windows necessary for more fabricating more complex components. In light of Figure 7, the manufacturers were establishing the knowledge-base for simple shapes, which provided the foundation for more complex parts such as the rotor. In contrast, U.S. firms were attempting to simultaneously optimize both material properties and processing conditions for the complex rotor itself. This proved a nearly insurmountable challenge, and many efforts were scaled back or discontinued altogether. Today, more than a decade later, only a few U.S. firms remain competitive in the advanced automotive component arena, and even then most of the market appears dominated by Japanese manufacturers.

INTEGRATED MODEL OF PRODUCT DEVELOPMENT - EXAMPLES

It is now useful to examine two cases where the integrated model of product development is being applied at Sandia National Laboratories. The first example describes a program known as "Smart-Weld"; the second example considers the compaction of ceramic powders.

(1) Case Study: "Smart-Weld"

The object of Smart-Weld is to integrate the areas of welding metallurgy, thermal analysis, stress analysis, and robot-controlled, automated laser-welding procedures into a unified, interactive software program for the product design engineer. The program begins by the design engineer creating a three-dimensional, CAD-based solid model of the proposed new design. From the material specifications of the design, a materials database is accessed, selecting the appropriate welding material and welding procedure. Knowing the constituent materials, information regarding mechanical and thermal properties is then obtained from a second database. This information is then integrated into the CAD-model, which is transferred to a module that calculates the thermal profiles required to form the weld. This module draws upon mathematical models that relate laser power input to temperature rise and fall within the desired weld region. The laser-weld operation is then simulated in another "virtual" welding module, that also establishes the appropriate computer code to drive the laser and robot in the actual welding process.

The next step in Smart-Weld is to ensure that the weld will have the adequate mechanical integrity. The simulated "welded" component in CAD form is relayed to a finite element module that automatically creates the FE mesh, and then estimates the stresses in the component. If the stresses are within an acceptable range (also determined by a set of rules within the module), then the go ahead is given for the actual laser weld. (If not, then the designer iterates the process with usually minor design changes until all criteria have been satisfied.) The final step in the process is to insert the two components into the welding fixture, and laser weld them together using a robot-controlled laser welder (recall the program for the robot was generated in an earlier module).

It is worth noting that the current version of Smart-Weld is a completely "point-and-click" software program operating from within a graphical user interface. The interface is merely a shell that contains the engineering modules that themselves contain the various informational databases. Thus the mathematical descriptions of the welding process, material properties, thermal analysis, etc., are completely transparent to the design engineer. The goal of the engineering development teams is to refine and update the databases, while the software links the databases together in a user-friendly fashion. Using Smart-Weld, new product designs require only one or two attempts to obtain the final welded component with the desired characteristics. This is a dramatic change from the weeks or months of trial and error that involved selecting appropriate material, varying the laser power and welding time, and testing the finished product for mechanical integrity. With this integrated model for product design, new weld designs are now possible within a few days of the original sketches.

(2) Case Study: Compaction of Ceramic Powders

In this example, the concept of an integrated approach to product development is applied to the compaction of spray-dried, ceramic powders. Recently, Sandia National Laboratories embarked on a program to develop experimentally validated, mathematical models of the compaction behavior of ceramic powders. These models ultimately would be linked together using the software interface developed for Smart-Weld to create a user-friendly, point-and-click program that could optimize the die design and material characteristics in a virtual manufacturing environment, as shown in Figure 8. Once optimized, the actual die could be fabricated to exacting specifications, and the material adjusted to produce powder compacts of uniform green density. Moreover, these compacts would require a single die to be machined, saving considerable time and expense.

SUMMARY

Important objectives of the advanced ceramic industry today are to reduce product development times and costs in order to improve competitiveness. One way of achieving these objectives is to minimize the number of design iterations and experiments necessary achieve the desired component shape and characteristics.

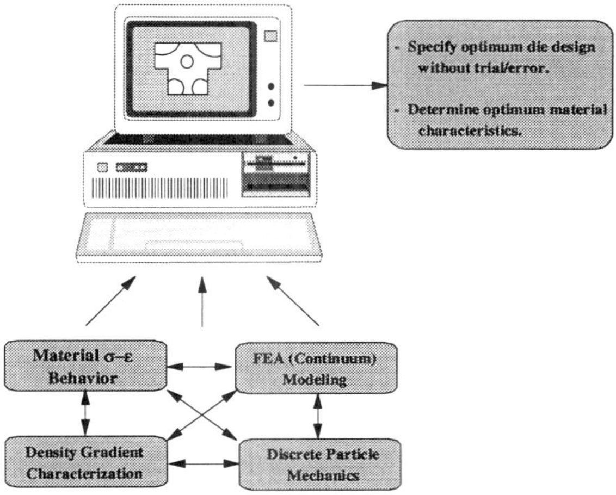

Figure 8. Integrated model applied to the compaction of ceramic powders.

By using a quantitative, integrated model of product development, design and manufacturing processes can be simulated in a virtual environment using today's desktop computer systems. Such simulation minimizes the number of iterations required to achieve the final component characteristics, and therefore greatly reduces development cost and time to delivery. Existing networking architecture provides the framework that facilitates the integration process, and also permits an expanding knowledge-base by capturing expertise attained in earlier projects.

ACKNOWLEDGMENTS

The author would like to thank Theresa Pollock and V. Arunachalam, both of Carnegie Mellon University for the many stimulating discussions that culminated in the ideas presented here. The author also acknowledges Kim Mahin for explaining Smart-Weld and Kevin Ewsuk and the Sandia Compaction Team for the many lively discourses on compaction. The author is also very grateful to Saundra Monroe for critically reviewing this manuscript. Financial support for this publication was provided by Sandia National Laboratories, operated for the U.S. Department of Energy under Contract Number DE-ACO4-94AL85000.

REFERENCES

References available on request.

AGILE MANUFACTURING, A SOLUTION FOR MANY CERAMIC INDUSTRY MANUFACTURING PROBLEMS

Charles L. Booth And Martin P. Harmer
Ben Franklin Technology Center, Northeast Tier And Materials Research Center, Lehigh University

ABSTRACT

There are many issues confronting the ceramics industry, particularly in the area of manufacturing. Several concepts or paradigms including Agile Manufacturing, Lean Manufacturing, Flexible Manufacturing, and Just In Time Delivery, to name a few, have been proposed as solutions leading to lower cost, rapid delivery, higher quality products. This paper will compare these "solutions" and discuss their advantages and disadvantages. The conclusion is reached that Agility, which is a system solution, in conjunction with electronic communication offers the best approach to solving many of the problems facing the ceramic industry today.

INTRODUCTION

As we move into the 21st Century all businesses must begin positioning themselves for new competition. Time to market with a constantly evolving mix of products and services will be the key to competitive advantage. This product mix delivery requirement puts a premium on cooperation among suppliers and customers. The ceramics industry is no exception. Customers will select vendors based on core competencies and skills and the ability to rapidly configure and reconfigure products to meet their evolving needs. The ceramics industry is not an exception to any of the following comments by business leaders [1].

"A key problem plaguing American business is that many organizations are vertically organized. In order to transfer this technology to the market place, one needs an integrated business team, i.e., an integration of functions, disciplines, and resources, -- a horizontal structure.

"In order to meet stringent customer demands in the new global market place, manufacturers are placing more emphasis on quality, time to market, and a host of other concerns. As manufacturers try to get all they can out of their production lines, they are finding that decision making needs to be moved downstream, close to the assembly process.

"Will it ever stop? Probably not ever- offshore competition that is. U. S. firms have had to improve productivity, quality, and distribution channels in order to survive the onslaught of global competition. The U.S. assembly industry has been challenged to support the growing number of offshore discrete manufacturing facilities...."

AGILITY

First a definition [2,3].

"Agile Manufacturing is the ability to respond very rapidly to unanticipated customer demands by putting together a variety of companies with different and complementary core competencies to make a new virtual corporation that exists as long as that market demand exists."

Agility then is the ability to thrive in a dynamic environment on a continuing basis. In an agile environment

- Products are customer opportunity based
- Customers will select vendors based on
  - core competencies and skills
  - the ability to rapidly configure and reconfigure products to meet evolving needs
- Organizations
  - will protect and grow their core skills
  - increase the strategic impact of people and information on the bottom line
- Virtual companies
  - will emerge to dynamically obtain the required skills for each product opportunity

Agility, a more general term than Agile Manufacturing, is accomplished by integrating all of the resources available to a company, i.e., technological and human, internal and external, customers, suppliers, and partners in joint ventures, into a coordinated, interdependent system. This system is capable of achieving short product development cycle times and responding quickly to sudden market opportunities. An agile company utilizes integrated processes for designing, manufacturing, marketing, and supporting its products and services. It's organizational structure is a customer-centered organization committed to sharply lower product development times and costs.

The four main principles of agility[4] are listed in Table I. In an agile environment, customers pay either a fee for skills, materials and a modest profit for products, or a percentage of the perceived value for solutions. Companies may adopt a value-based strategy to configure products and services into solutions which enrich their customers.

Table I. Four Principles of Agility

1. Enriching customers: Products vs. solutions
2. Cooperating to enhance competitiveness: Virtual Organizations
3. Mastering change and uncertainty: Entrepreneurial Organizations
4. Leveraging: Impact of people and information.

In an agile organization, cooperation enhances competitive capability. Companies use the virtual company model internally as well as externally to share responsibility and enhance cooperation opportunistically across organizational lines. Agile competition is based on the ability to thrive on change and uncertainty. Agility's entrepreneurial organization responds more quickly than a hierarchical structure to changing conditions. In an agile environment, organizations sell skills, knowledge and information over time. Companies make investments to increase the strategic impact of their people and information on the bottom line.

MANUFACTURING OPERATING ARCHITECTURES

Several concepts or paradigms including Agile Manufacturing, Mass Manufacturing, Lean Manufacturing, Flexible Manufacturing, and Just In Time Delivery, to name a few, have been proposed as solutions leading to lower cost, rapid delivery, higher quality products, i.e., solutions for existing and evolving business and manufacturing needs.

Several manufacturing configurations [5] are presented in Figure 1. At one extreme is the original craftsman who had comprehensive knowledge of one skill or technique but virtually no knowledge of other areas. These craftsman still exist today. Examples are tool and die makers. In a ceramics plant they might be the mold maker.

Operating Architectures

	Craft	Mass	Lean	Agile
Reconfigurable				
Flexible				
Fixed		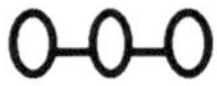		
Comprehensive				

Figure 1. Operating Architectures

As manufacturing demands became more complex, Mass Manufacturing techniques were applied. In this configuration there is a linear organization with fixed lines of communication. Several characteristics [6] of this configuration are listed in Table II. People are still experts in one area but the various areas are coordinated to increase manufacturing capacity. Any movement of people, product, or information between areas are strictly along well defined rigid lines. The manufacturing organization and factory floor are designed to make a large number of one item with no changes allowed. High investment costs are amortized over the large numbers of pieces manufactured on the expensive equipment.

<u>Table II. Mass Manufacturing</u>

- Invest $$ In Equipment & Facilities
- High Touch Labor
- Specialization
- Little Worker Input
- Many Layers
- Slow Decision Making
- Workers Could Buy Products They Made
- Product Development Time In Years
- Inconsistent Quality

Lean Manufacturing [6] in one sense was the next innovation, Table III. The areas of expertise are still maintained with rigid lines of communication and some reconfigurability between units. Information and personnel are retained in individual units with communication along well defined lines and protocols. Lean manufacturing tries to remove all waste from the system with a minimum inventory utilizing Just In Time inventory principles. Fixed teams are in place to reduce cycle time as market pull develops for the product and there is some flexibility designed into the manufacturing process for anticipated changes.

Table III. Lean Manufacturing

- Invest $$ In Technology
- Removed Layers
- Reduced Waste
- Worker Input
- Established Teams
- Greater Dependence On Suppliers
- Stable Process
- Product Development Time In Months
- High Quality At Point Of Sale

Flexible manufacturing in one sense is a feature of Lean Manufacturing in that it allows the system to manufacture a variety of products in a cost-effective, timely manner. The manufacturing process has built in planned adaptability and is reconfigurable for low volume manufacture of a high quality product. Rapid response is possible within the design guidelines of the process.

Agile Manufacturing, the other extreme from the individual craftsman would appear to be just a further extension or progression in the series from craftsman to Mass Manufacturing to Lean Manufacturing. In one sense it is. However, there is an important difference in that the rigidity of the other systems is gone and there is complete overlap of areas of expertise without fixed lines of communication. Agile Manufacturing manages unexpected or unplanned changes in the manufacturing process resulting from customer driven needs. Some of the characteristics of Agile Manufacturing [6] are presented in Table IV. One of the primary differences in Agile Manufacturing is the scope of permitted changes, i.e., Flexible Manufacturing limits changes to planned changes while Agile Manufacturing encourages unlimited, unplanned changes. However, flexibility is one key element of agility.

Table IV. Agile Manufacturing

- Invest $$ In People & Information Systems
- Multi-skilled Work Force
- Robust, Reconfigurable Teams,Equipment, Infrastructure
- Partnering In All Stages Of Manufacturing
- Self-directed Work Teams, Management Break Barriers
- Effective Technology And Information Integration
- Development Time In Weeks
- High Quality Across Product Life

The differences in how Lean and Agile utilize the various manufacturing units such as personnel, suppliers, scheduling are compared in Table V. People, their skills, and information are the most important organizational asset of an agile organization.

Table V. Lean / Agile Manufacturing Comparison

Element	Lean	Agile
Assets	Technology	People & Info.
Workers	Input	Multi-skilled
Teams	Established	Reconfigurable
Suppliers	Dependence	Partners
Delivery	Months	Weeks
Production	Flexible	Reconfigurable

Agile organizations form partnerships between users and suppliers, both realizing that secrecy between users and suppliers results in equipment and materials that lag process designs, ultimately having a negative impact on cost, performance, and customer satisfaction. Agility takes the team concept a step farther in that teams are reconfigurable to meet specific needs as long as the need exists. However, for these teams to succeed, they must truly be empowered by management and have adequate training in both computer and communication skills. Companies are required to make decisions faster and closer to the source of information. Self-directed, empowered work teams are a requirement for success.

VIRTUAL CORPORATIONS

The competitive environment that is pulling Agile Manufacturing into existence is characterized by the apparent paradox of intensifying competition among producers and intensifying cooperation among them.

The formation of virtual corporations [7] involves the dynamic partnering of firms to quickly combine core competencies to attack market opportunities. In a virtual company each partner is chosen to provide a unique capability and the number of partners may vary over time as needs dictate. The virtual corporation is founded on the principle that it can prosper more by cooperation and sharing than it could by going it alone. The partners dynamically manage and share as appropriate

- Cost and revenues
- Personnel
- Knowledge, skills, and intellectual property
- Customer opportunities and access to global markets

The characteristics of a virtual corporation are given in Table VI.

<u>Table VI. Characteristics Of The Virtual Corporation</u>

<u>The Ultimate In Adaptability</u>

- Opportunism
- Excellence
- Technology
- No Borders
- Trust

In this agile virtual environment, products are customer opportunity based. Customers select vendors based on core competencies and skills and the ability to rapidly configure and reconfigure products to meet their evolving needs. Vendors will form virtual corporations to meet these customer demands with an emphasis on horizontal integration between companies rather than vertical integration within single companies. The horizontal partnerships will enable the group to rapidly obtain the required skills to produce products beyond the capability of the individual corporations.

The virtual corporation is only an association of convenience and will change membership with time as product demands change. This dynamic arrangement will increase the strategic impact of people and information on the bottom line with its ability to rapidly configure and reconfigure products meeting evolving needs. These virtual companies will emerge to dynamically obtain the required skills as each product opportunity develops.

Within the virtual corporation each organization will protect and grow their core skills while contributing to the success of the whole. Each member retains ownership of intellectual property it had prior to association but shares equally in everything developed by the virtual organization. Among the obstacles to rapid formation of virtual companies are some thorny legal issues (including the protection of intellectual property rights), as well as finding ways to determine the value of each participants contribution. There are cultural issues, such as the tendency of American companies toward autonomy and a reluctance to share proprietary information. More to the point: a tendency toward distrust. People must be willing to share data and work with someone who has been perceived as a competitor.

AGILE WEB

An agile organization can be characterized by an agile web arrangement, depicted in Figure 2. The diagram illustrates the flow of information and decision making from the initial customer inquiry to product shipping and subsequent follow up. Agile principles are applied at all stages. When the Web configures a team, the team might involve personnel from an outside firm with necessary core competencies, forming a virtual company to facilitate planning and ultimately decrease delivery time. Teams will cross functional department lines, eliminating colloquial thinking and narrow ideas. The team might comprise design, manufacturing, marketing, purchasing, finance, customers, suppliers, and even competitors which have a core competency required for success. This team must have the authority to assemble the resources required for success. The solution proposed by this team is then the best solution not of the initial company, but the best solution possible from the virtual company which has maximized all of the core requirements to produce the best product in the shortest time. As the procedure moves forward, additional firms can be recruited and added to the virtual group as needed.

The entire process is designed to eliminate, not minimize roadblocks, in information flow as well as technical matters. Adequate, correct information flow between all participants is essential. Information will be shared electronically and all tooling designs downloaded electronically for accuracy. Information flow will achieve knowledge flow not simply data flow. In a non-agile environment, technical delays are accepted as a matter of course in any new development and information delays in many instances are not even recognized. In a non-agile environment information moves via paper through an internal mail system possibly requiring 5 days to move two offices away from the originator. Each participant is concerned with meeting his deadline and builds in a safety factor to insure he will meet his target, forgetting that the customer's requested shipping date is the important factor, not an employee's personal record of meeting targets. In an Agile organization information moves electronically and everyone has a level of trust in each other. Credibility is easier to establish when the team was assembled with core competencies in mind rather than "who is available with nothing to do." This level of trust and competence is an important factor in achieving realistic time estimates.

Agile Web

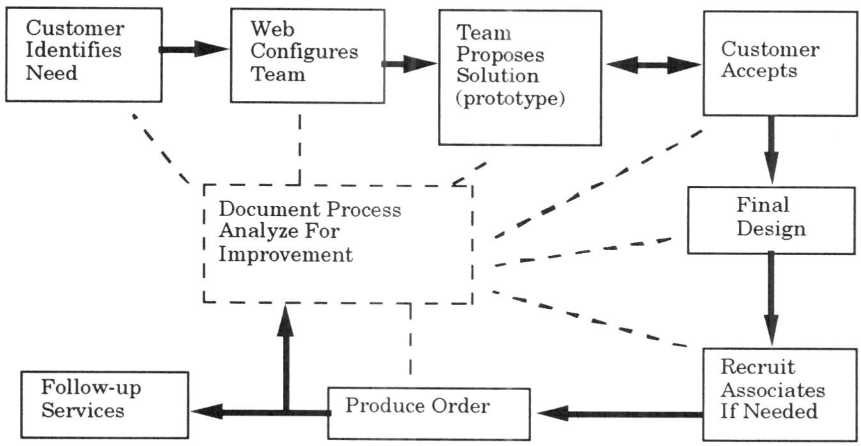

Figure 2. An example of an Agile Web Manufacturing system

AGILE CERAMICS

The initial reaction by many ceramists is that the ceramic process is too variable, consisting of many hand, custom operations on small quantities that do not lend themselves economically to automatic production. This type product mix is ideal for agile principles. Figure 3 represents the steps in a typical sequence in placing an order for a new ceramic part [8]. It's details fit nicely into the boxes of Figure 2. Cooperation between trusting partners, utilizing electronic inter-and intra-company communication, and sharing people, material resources, data would substantially impact delivery times, cost, and quality.

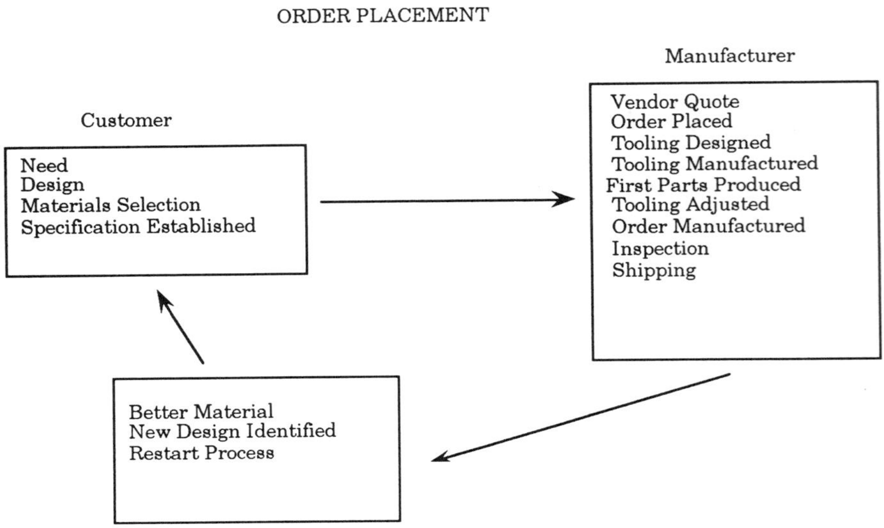

Figure 3. Typical Order Placement Sequence

Historically the ceramics industry has not been agile in its thinking about product solutions to customer problems.

- "You want aluminum nitride? We don't manufacture aluminum nitride, sorry."
- "We will only manufacture zirconia toughened alumina from now on since there is no need for other alumina."
- "New tooling is required. Delivery time for tooling is 8 weeks and will cost $$$$."

Many more of these phrases exist. Their existence and the lack of agility in ceramic companies is illustrated by the increasing number of non-ceramic companies that are establishing in-house ceramic manufacturing capability. Lack of response of existing manufacturers is given as the reason. Agility would provide these companies the opportunity to form virtual companies and the ability to meet their needs with existing suppliers without the permanency of a new internal division. If existing companies were more agile, these firms would not consider going captive since manufacturers would have formed virtual webs to enhance their capability and meet the needs of these potential customers. An agile web would permit manufacturers without aluminum nitride capability, for instance, to participate in this business by partnering with other manufacturers. With pooled resources all would benefit and shared resources such as tooling would be transparent to the customer while benefiting everyone.

Most ceramic manufacturing processes involve tooling of one kind or another and setup time can be time consuming and limit productivity. Many times the tooling is only needed for a few parts and then must be replaced with minimal down time.

While not the ceramics industry, a John Deere equipment manufacturing division faced the same problem and solved it with Agile Manufacturing [9]. Their approach can be applied to ceramic operations.

John Deere Example:

In one of John Deere's manufacturing operations they were pushing the envelope of stamping press time. The objective was to stamp a variety of farm implement components which were currently being done by contract stampers, and do it in-house on a refurbished press. The trick was to do this without spending more time changing dies than stamping parts. The system designed to achieve rapid die change included

- Powered unloading and loading of the dies into the press.
- Automated integrated movement of dies to and from close-proximity storage and in and out of the press. The system functions somewhat like a tool changer on a machining center.
- A multi tier storage rack unit with individual compartments for up to 20 dies that are accessed by an elevating die cart. The system allows storing and retrieving dies during production or die change without disrupting either. The racks also can store stacks of steel blanks.
- An Atlas T-table which uses a powered push/pull module to move dies into and out of the press. The T-table has two temporary storage positions for dies that have just been removed or are upcoming prestaged dies just delivered by the die cart.

The result was a 2.5 minute die-to-die change over for a system averaging five changeovers per day for batch runs that are under 100 parts each. There are many ceramic plants in which a 2.5 minute die-to-die change over would be a 10X improvement.

SUMMARY

Over time, manufacturing systems have evolved from the original craft system through Mass Manufacturing to Lean Manufacturing. Mass and Lean Manufacturing retained the original craft specialities but coordinated them to perform repetitive operations in a high volume low cost environment. Each step in these processes involved a "craftsman," expert in his one area but having no familiarity with and limited input to other areas. Investment in technology and equipment was emphasized with little regard for personnel and communication. These traditional organizations utilized linearly ordered tasks with sequential information passed in a preferred direction, i.e., downstream for conveying assignments and specifications and upstream for reporting problem situations and progress.

The progression of manufacturing organizations from the original craft system through Mass and Lean Manufacturing has now lead to Agility or Agile Manufacturing. However, the progression from previous systems to agility is not linear since it is a new paradigm. Agility addresses information flow, importance of people, cooperation, and flexibility, and will utilize the attributes of Lean and Mass Manufacturing, TQM, etc. where appropriate to meet a customers needs. Today business relationships can take longer to develop than the technical content of a product. An agile environment will dramatically increase the quality of the relationships and shorten the time for forming them.

Within this new paradigm customers select vendors based on core competencies and skills and obtain the ability to rapidly meet new product needs. Virtual corporations or webs are formed to enhance response to customer needs by sharing costs, personnel, facilities, and equipment. These relationships are temporary, lasting only long enough to satisfy a specific need. The increased flexibility offered by these arrangements provides a means for companies to expand their product offering without in-house investment. Small companies appear to be "large" in capability while large corporations appear "small" in response time and flexibility.

Agility cannot be purchased and its implementation is different for each company. Becoming agile requires

• Changing policies and procedures
• Identifying and removing barriers
• Eliminating non-value adding processes
• Mutual trust within and between cooperating partners

An example has been described where one company, John Deere, substantially improved its production operation using Agile Manufacturing principles. Application of this new "agile paradigm " in the ceramic industry should begin limiting the number of companies becoming captive manufacturers due to the lack of customer response from the traditional suppliers.

REFERENCES

1. "Challenges Continue for U.S. Advanced Ceramics", L. M. Sheppard, Ceramic Industry, May 1994, p. 36ff.

2. "Manufacturing Agility: The 21st Century Strategy", M. Weil, Managing Automation, August 1994, p. 18ff.

3. Agile Competitors and Virtual Organizations, Goldman, Nagel, Preiss, Van Nostrand Reinhold, 1995.

4. "Understanding Agile Competition, A Quick look at How to Make your Company Agile", Roger Nagel, Iacocca Institute, Lehigh University, 1993.

5. "Tools for analyzing and Constructing Agility", R. Dove, Agility Forum, Bethlehem, PA, 1994.

6. "What is Agility", C. Ward, Business Strategies, 1995.

7. "Agility, the Future for Ceramic Manufacturing", C. L. Booth & M. P. Harmer, Ceram. Eng. Sci. Proc., **16**[1] 220-25(1995).

8. "Agility, A Concept for the 21st Century", R. Patterson, Iacocca Institute, Lehigh University, Presented at the 1994 Fall Meeting of the Materials & Equipment, Whitewares, Refractory Ceramics, and Basic Science Divisions, Louisville, September 1994.

9. "The Search for Agile Manufacturing", P.M. Noaker, Manufacturing Engineering, November 1994.

REDUCING PIM DESIGN-TO-MANUFACTURE CYCLE TIME WITH PROCESS MODELING AND DIE DESIGN EXPERT SYSTEM

K. E. Carr*, C. M. Wang, K. J. Meltsner and T. J. McCabe, Concurrent Technologies Corporation, Johnstown, PA, 15904

ABSTRACT

Powder Injection Molding (PIM) is a net shape process for manufacturing both ceramic and metal components. PIM is used for cost-effective mass production of small complex-shaped components that are difficult to form or machine by other methods. However, extensive trial-and-error procedures are usually necessary to determine the proper die design and molding conditions. This paper describes the development of process simulation and expert system design tools to reduce the design-to-manufacture cycle time and improve component quality. The effect of ingate and vent location and process parameters is predicted with the PIM mold filling process simulation code. This 3-D numerical analysis code predicts defects, such as voids, sink marks, weld lines, density variations and short shots that result if molding parameters and tooling features are not properly specified. The die design expert system combines a commercial CAD package and an expert system to create a mold cavity design tool. With the feature-based expert system knowledgebase, the original part dimensions are selectively expanded to form the mold cavity and compensate for non-uniform shrinkage.

INTRODUCTION

Powder Injection Molding (PIM) is a net shape manufacturing process used for cost-effective mass production of small complex-shaped components that are difficult to form or machine by other methods. However, extensive trial-and-error procedures are usually necessary to determine the proper die design and molding conditions. With ever increasing marketplace competition and the drive to implement agile manufacturing and concurrent engineering concepts, long lead times to develop tooling and manufacturing parameters are increasingly unacceptable. Process simulation, expert systems, and virtual manufacturing are becoming viable options for today's designers to increase the speed with which a product enters the marketplace.

To address the need for these tools for PIM, the National Center for Excellence in Metalworking Technology (NCEMT) is developing the PIM Mold Filling Process Simulation Code and Die Design Expert System to reduce design-to-manufacture cycle time. With the PIM Mold Filling Process Simulation Code (PIM Code), the effects of gate location, venting, temperature, and pressure can be determined. A simulation can be completed in a matter of hours, which allows several variations to be considered before tooling is designed. With the Die Design Expert System, designers will have an expert system to aid them in compensating for non-uniform shrinkage from molding, debinding, and sintering.

By using the PIM Code and Expert System, a designer will be able to refine the location of mold features such as ingates and vents. Then the designer will modify the original part geometry to form the mold cavity, thus compensating for dimensional changes which occur during the debinding and sintering cycles. Use of these tools will reduce rework on dies and reduce the need for iteratve tooling creation, which in turn will save time and money.

At present these tools are intended for metal powder injection molding, but both are easily adaptable to ceramic injection molding through the use of different material parameters. In the case of the mold filling simulation, ceramic and metal injection molding are very similar; the greatest effect on material properties comes from the loading fraction and the polymer and wax binder system. With the Expert System, compensation rules will be developed from empirical studies and calibrated to different materials.

PIM CODE

The PIM Code was developed to aid the designer in locating vents and ingates and predicting whether or not the design is producible. With knowledge of the location of potential defects, a designer may redesign the mold to either remove defects or relocate them into non-critical areas. The PIM Code also predicts potential molding problems arising from machine settings such as injection and mold temperature, injection pressure, and mold heating or cooling.

Using a 3-dimensional color graphics program, the PIM Code can display both primary and derived information from the simulation. Primary output information "direct display" includes flow front locations and velocity, temperature, and pressure distribution, all at any time during the mold filling sequence. Derived information includes shear rates, shear stresses, formation of weld lines, and occurrence of short shots. Through pressure calculations, the PIM Code can also predict the occurrence of voids and short shots that arise from inadequate venting.

This process simulation code differs from commercial plastic injection molding codes in that it is based on the implementation of a fully 3-dimensional fluid flow and heat transfer solution, designed especially for the highly loaded suspensions found in PIM. Thicker cross sections and arbitrary 3-dimensional geometries can be more effectively modeled by the PIM Code than by plastic injection molding codes based on the generalized 2-dimensional Hele-Shaw model for thin cavities.[1] The PIM Code employs the finite difference method to implicitly solve the partial differential equations describing fluid flow and heat transfer, while taking into account viscous heating phenomena and time-dependent non-Newtonian fluid flow behavior with free surfaces. The volume of fluid method is applied to track the melt front advancement.[2] Details of the mathematical technique have been presented elsewhere.[1,3]

In order to approximate the mold cavity, a uniform shrinkage factor is applied to the original part geometry. Because detail is lost in building a finite difference mesh, the PIM Code is not sensitive to the small dimensional changes that a non-uniform shrinkage factor would provide. Sensitivity to detail improves by using a finer mesh, but this increases computation time and is rarely needed for an adequate solution.

Verification is critical before confidence can be placed in any simulation. Previous papers have compared PIM Code simulation results to metal injection molded short shots.[1,4] The results have been very similar, providing confidence in the ability of the model to predict flow front advancement.

Additional metal injection molding experiments were used to further verify the model. The results of the PIM Code mold filling prediction are presented below in Figures 1-5, and compared with the actual short shots. As can be seen, the PIM Code accurately predicts both the uneven filling of the legs and the sequence of mold filling.

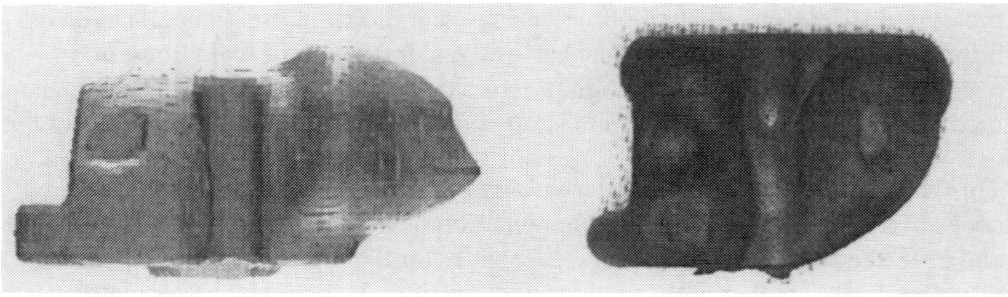

Figure 1. Comparison of PIM Code simulation result with actual short shot. Simulation results are on the left.

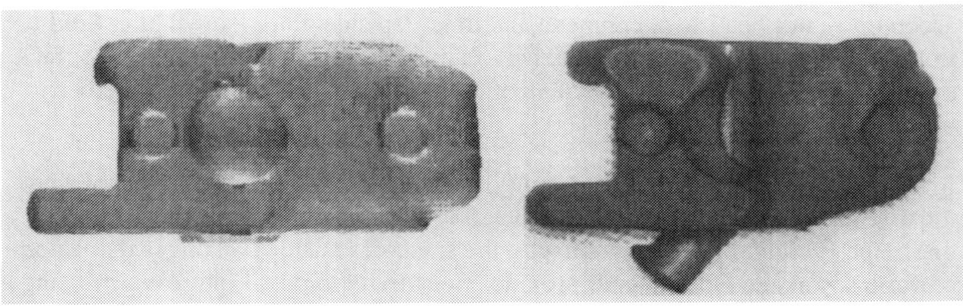

Figure 2. Comparison of PIM Code simulation result with actual short shot. Simulation results are on the left.

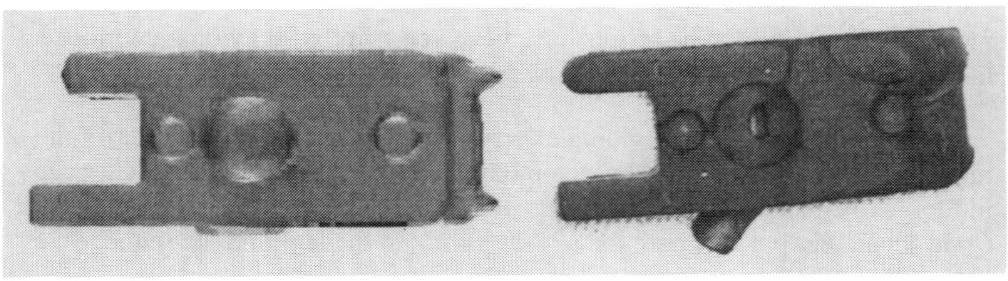

Figure 3. Comparison of PIM Code simulation result with actual short shot. Simulation results are on the left.

Figure 4. Comparison of PIM Code simulation result with actual short shot. Simulation results are on the left.

Figure 5. Comparison of PIM Code simulation result with actual short shot. Simulation results are on the top.

When trying to predict production rates from molding cycle times, process planners and cost estimators find knowledge of the cooling profile to be useful. Figure 6 shows output from the PIM Code in a cutaway view at two different times. The ability to predict the internal temperature allows decisions to be made regarding ejection timing. Ejecting a part before the interior has cooled to the point of solidification will contribute to residual stresses and distortions.

In addition to the flow fronts, the temperature profile may also be used to predict weld lines. When two flow fronts meet, a weld line may form when the temperature of the feedstock is such that inadequate mixing occurs. The PIM Code can alert the designer when the meeting flow fronts are below the temperature that will allow the material to combine.

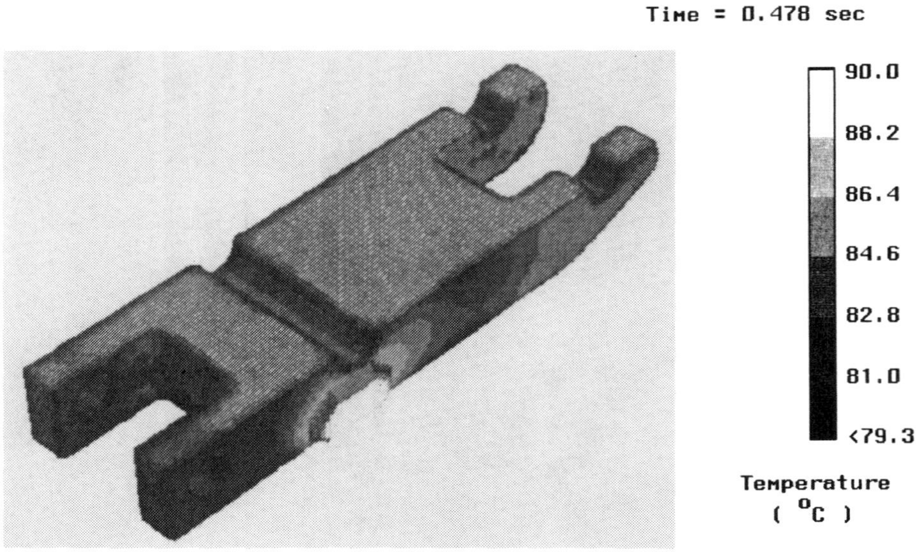

(a)

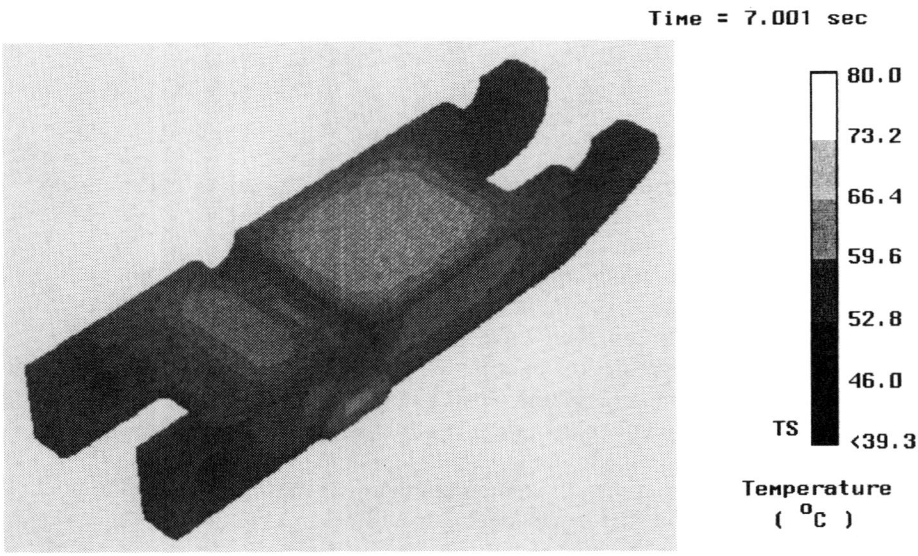

(b)

Figure 6. PIM Code simulation results showing temperature distribution over time; (a) 0.478 seconds (end of fill) and (b) 7.001 seconds.

The PIM Code can also display velocity profiles. These can be displayed as velocity vectors in any plane of the simulation. An example of velocity vector output is shown in Figure 7. Velocity vectors can be used to predict the alignment of whisker reinforcement, thus predicting local mechanical properties in composite materials made by PIM.

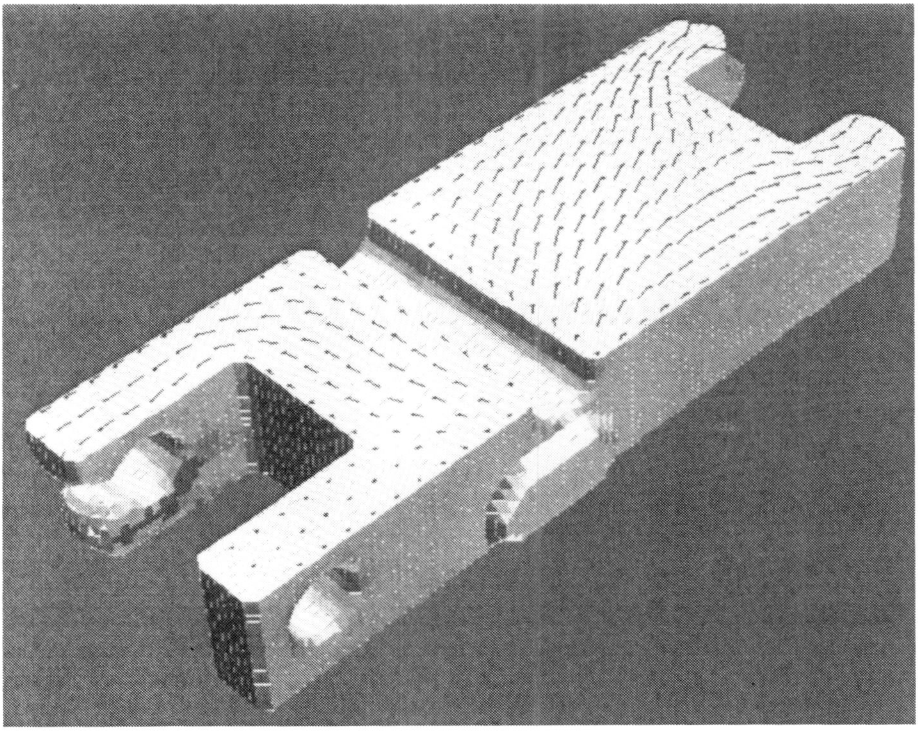

Figure 7. PIM Code simulation result showing velocity vectors during filling.

The information provided by the PIM Code can be used to aid the designer in optimizing the mold and injection conditions, and for predicting cycle times. If a problem is found with the producibility of the original part design, modification of the design may more easily be made at this stage. Once the part geometry and the mold gating and venting have been defined, the Die Design Expert System can be used to modify the part geometry to control the dimensions of the finished part.

EXPERT SYSTEM

The Die Design Expert System aids the PIM designer by modifying the part geometry to compensate for non-uniform shrinkage factors. Once the PIM Code has been used to define a moldable design and appropriate mold gating and venting, flow direction can be determined. This information is needed to predict non-linear shrinkage which arises from flow direction, powder particle shape, feedstock rheology, section thickness, non-uniform powder distribution, and mechanical constraints from part features. These factors can be broken down into a dependence on injection direction, local section thickness, and part features.

Although non-linear shrinkage is known to occur, little more than anecdotal evidence exists and optimization rules cannot be derived from basic principles. Thus, an empirical approach will be used for the expert system instead of approaching the problem through simulation. Using sintered part and original die cavity measurements, correlations will be made and general rules developed. Special attention will be paid to different features such as ribs and holes to determine their effect on non-linear shrinkage. This process must be repeated for various feedstocks. Eventually experience will be accumulated across part families and different feedstocks.

An advantage of using an expert system is that deep knowledge and understanding are not required at first. For the system to be successful, only general rules must be developed. With experience and a broader knowledgebase, any universal compensation rules that exist will be uncovered. Unfortunately, without the appropriate input, the system will not handle new features or feedstocks.

The Die Design Expert System is built on Stone & Webster's StoneRule software to interface with Parametric Technology Corporation's Pro/Engineer. The expert system can analyze potential origins of non-linear shrinkage, then correlate them with the part design. Based on compensation rules related to the local section thickness and injection flow direction, the original part dimensions are then adjusted. An example of a general compensation rule is as follows:

IF

  section thickness < 1 cm and > 0.5 cm

  flow direction is perpendicular to dimensions

THEN

  final dimension = desired dimension * 1.26

Part geometry expansion to form the mold cavity can be performed using Pro/Engineer's parametric design capability.

Work is still in progress on developing the Die Design Expert System. Several PIM companies are providing the necessary parts measurements to build the expert system knowledgebase. At this time, there are no plans to factor out the molding, debinding, and sintering stages as separate areas that cause dimension alteration. When work is completed on the Expert System, the PIM Code and Expert System will provide a powerful set of tools for the PIM designer.

SUMMARY

In the traditional design methodology, a part designer optimizes a part for an application without concern for whether the design is producible. The part design is then "thrown over the wall" to a PIM designer who must adapt the original design to make it manufacturable by PIM. The adapted design is then received by a die designer who is responsible for designing a die for a part that may still be, at best, marginally producible. Finally, the die maker builds the die, usually with the cavity somewhat smaller than required to allow reworking of the die to produce the desired geometry. Furthermore, part dimensions may vary in production with changes in powder loading in the feedstock or other processing variables. This process, shown schematically in Figure 8a, is time-consuming, inefficient, and not conducive to good parts production.

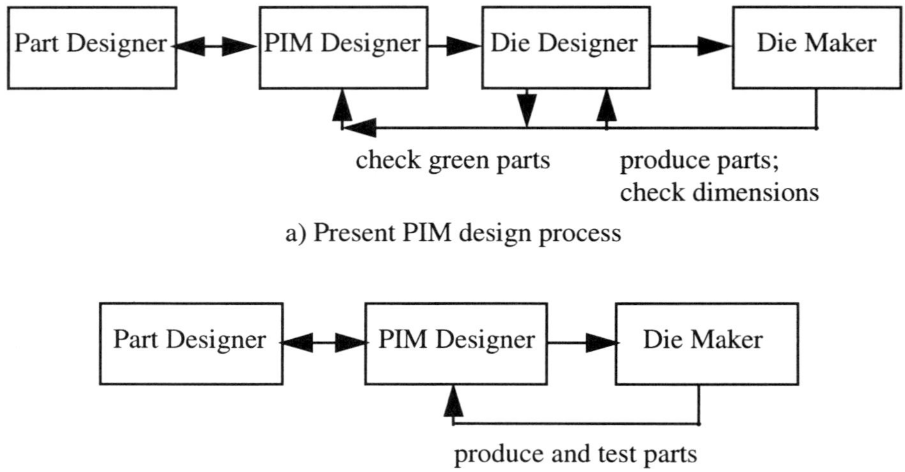

a) Present PIM design process

b) Proposed PIM design process with PIM design tools

Figure 8. Schematic flow of the PIM design process.

With process simulation tools and expert systems, such as the PIM Code and the Die Design Expert System, the PIM designer is given the information necessary to make proper decisions in the original process design stage. This encourages concurrent part and process design, thus ensuring that parts are designed for manufacturability by PIM, not force fit from other manufacturing processes. This process, shown in Figure 8b, speeds design, increases efficiency, and provides a better product.

As with most manufacturing operations, the earlier potential manufacturing problems are brought into the design cycle, the less they will cost. Providing the tools for concurrent design will save effort and money later in the cycle. PIM molds are generally very expensive, with long lead times to manufacture. While die rework can accommodate many dimensional fixes, mistakes in threads, hole placement, or other fixed features may require a new die to be produced at a significantly higher cost and delay.

These tools will allow virtual part production and expert system assistance in applying shrinkage factors that should eliminate much of the trial-and-error involved in die design. Reduced trial-and-error means reduced cost and decreased design-to-manufacture cycle time. With a suite of tools including the PIM Code and Die Design Expert System, the NCEMT is providing the PIM community with a method to increase agility and reduce design-to-manufacture cycle time.

ACKNOWLEDGMENT

This work was conducted by the National Center for Excellence in Metalworking Technology, operated by Concurrent Technologies Corporation under contract to the U.S. Navy as part of the U.S. Navy Manufacturing Science and Technology Program.

CTC would like to thank R. K. Viswanadham of Vermont American Multi-Metals Division for providing the short shot photos and information. In addition, the authors would like to thank Shawn Nesmith for preparing figures from the PIM Code output.

REFERENCES

1. C. M. Wang, R. L. Leonard, R. A. Posteraro and T. J. McCabe, "A Finite-Difference Computing Model for 3-D Simulation of the Powder Injection Molding Process," *Proceedings of the 1992 Powder Injection Molding Symposium*, edited by P. H. Booker, J. Gaspervich, and R. M. German, Metal Powder Industries Federation and American Powder Metallurgy Institute, Princeton, NJ, 1992, pp. 435-449.

2. C. M. Wang, R. L. Leonard and T. J. McCabe, "Prediction of Defect Formation in Powder Injection Molding by Computer Simulation Technique," Third International Conference on Powder Metallurgy in Aerospace, Defense and Demanding Applications, San Diego, CA, February 8-10, 1993.

3. C. M. Wang, R. L. Leonard and T. J. McCabe, "Computational Analysis of Fluid Flow and Heat Transfer in the Powder Injection Molding Process," *Advances in Powder Metallurgy & Particulate Materials - 1993, Vol. 5*, compiled by A. Lawley and A. Swanson, Metal Powder Industries Federation and American Powder Metallurgy Institute, Princeton, NJ, 1993.

4. C. M. Wang, K. E. Carr and T. J. McCabe, "Computer Simulation of the Powder Injection Molding Process," *Advances in Powder Metallurgy & Particulate Materials - 1994, Vol. 4*, compiled by C. Lall and A. J. Neupaver, Metal Powder Industries Federation and American Powder Metallurgy Institute, Princeton, NJ, 1995.

DUCTILE REGIME MACHINING OF BRITTLE MATERIALS

R. O. Scattergood, Department of Materials Science and Engineering, North Carolina State University, Raleigh, NC 27695-7907

INTRODUCTION

The machining of brittle materials is an important issue for the manufacturing of components from these materials. Grinding of advanced ceramics can in fact contribute 50 % or more to the total cost of a finished component even with near net-shape production techniques. Control of the subsurface damage is one of the crucial factors for the machining operation since subsurface damage can severely limit the performance of brittle materials. With the advent of high-stiffness, high-precision machine tools it is now possible to machine brittle materials without introducing subsurface fracture damage. This limit has been termed "ductile-regime" machining. The profile of the uncut chip in a machining operation is central to the understanding of the subsurface damage produced and to the concept of ductile-regime machining. Chip profile geometry will interplay with the material removal mechanisms such that the damage left in the part will be controlled by both the geometry and the material response. The latter is best characterized as a critical depth-of-cut for the onset of a ductile-to-brittle fracture transition. Also important is the size scale of the fracture damage produce at this transition. Material microstructure will play an important role in determining the material response, especially with toughened ceramics. The factors underlying the ductile-regime machining concept and the production of subsurface damage during a machining operation will reviewed in this paper. Single-point machining will be used as a prototype because the geometrical aspects are most easily understood in this case and the concepts have been verified. The conceptual framework will then be extended to grinding using plunge grinding as a prototype. This introduces important modifications because the chip profile geometry can be complex for a grinding operation and the cutting tool is made up of a large number of active cutting grains as opposed to a single cutting point. Since many of the viewpoints introduced here differ from the more conventional views of machining processes and concomitant fracture damage, a significant part of the discussion appears as a synopsis of current research efforts.

SINGLE POINT MACHINING

The essential connections between chip profile geometry, material removal mechanisms and the fracture damage introduced into a workpiece can be illustrated using a single-point machining operation. For brittle materials, this type of machining is currently used to fabricate high-precision components for applications such as optical systems and infrared detectors. Since single-crystal diamond tools are needed to obtain the surface finish and form accuracy required, the process is usually called "single-point diamond turning". High-precision computer-controlled lathes with positioning accuracy the order of a few nanometers are required. It must be noted, however, that single-point diamond turning, although quite successful for certain classes of brittle materials, has limited applicability in manufacturing. Grinding is a much more versatile process and will be the choice for most higher volume manufacturing applications. This is the primary motivation for the development of a process model for the grinding of brittle materials.

Fig. 1 shows the geometry for a simple facing cut made with a round-nose single-point diamond cutting tool. Due to the cutting motion and feed motion (crossfeed) of the tool, the uncut chip profile appears as the shaded section shown in Fig. 1. It is important to note that this geometry differs from the orthogonal cutting geometry that is widely used for describing machining processes [1].

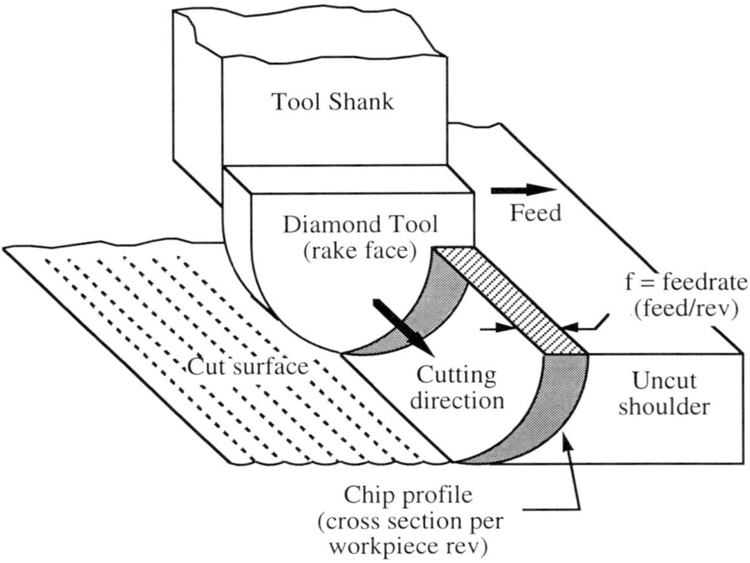

Figure 1. Single-point diamond turning used a round-nose cutting tool.

The effect of material response and its interplay with chip geometry can be illustrated using Fig. 2, which is a projection of Fig. 1 along the cutting direction. The uncut chip profile is the sector bounded by two circular arcs whose radius is the tool nose radius. Separation of the arcs is determined by the feed-per-revolution f, which is termed the feedrate. The effective depth-of-cut d varies from zero at the finished surface to a maximum value at the top of the uncut shoulder. For brittle materials, the extensive literature on indentation fracture [2] shows that there must be a transition from ductile plastic flow to microfracture at a critical depth-of-cut d_c. This can also be expressed as a critical load P_c, however, d_c is preferred since machining operations inherently control the cutting depth rather than the load. Fracture damage with a characteristic size scale y_c will be initiated at d_c as shown in Fig. 2. This "ductile-to-brittle" transition is endemic to any machining or grinding process used for brittle materials like glass, semiconductors or ceramics. The interplay between chip profile geometry and material response follows directly from Fig. 2. If the depth of fracture damage y_c extends below the plane of the finished surface (cut surface plane), then the part will contain subsurface damage and its extent will depend upon the location of d_c along the chip profile boundary. The relevant machining parameter is the feedrate f; smaller values of f will move d_c higher along the chip profile. There will be maximum value f_{max} such that for $f < f_{max}$ the finished surface will remain damage free even though a significant amount of material removal can still occur by fracture in the uncut shoulder. This damage is removed by subsequent passes of the cutting tool. In the classical orthogonal machining model [1], the total depth-of-cut h would be used in place of the effective depth-of-cut d in Fig. 2. In essence, single-point machining with a round-nose cutting tool is a sequence of orthogonal machining cuts with a variable depth-of-cut d. It is very important to recognize the effects of chip profile geometry here. Changes in the chip profile due to changes in parameters such as feedrate f will control the fracture damage for given material response parameters d_c and y_c. These latter parameters will vary with material properties and tool loading conditions. High precision machines using air bearings are required for achieving ductile-regime machining in practice since the critical feedrates f_{max} will typically be the order of a few micrometers per revolution. Machines with conventional bearings and lower stiffness will have error motions in excess of a few micrometers and so cannot maintain adequate feedrate control, i.e., the uncontrolled variations Δf in feedrate f will be sufficient to generate subsurface damage.

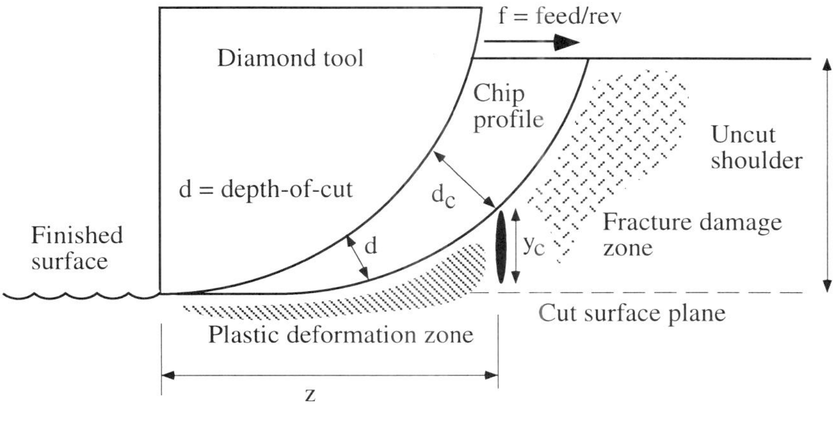

Figure 2. Projection of single-point diamond turning along the cutting direction. For typical conditions, the distance h is very much less than the horizontal extent of the chip profile (this is distorted in the figure for clarity).

Interrupted cutting tests for single point diamond turning have been made to verify the influence of the chip profile geometry shown in Fig. 2. This test is relatively easy to implement using a PZT servo actuator fixed in the tool holder. This servo rapidly retracts the tool point from the workpiece. Most of the results have been obtained using germanium single crystals. Single crystal material avoids complexities due to microstructure and provides more easily diagnosed test results. Silicon can also be used, but the wear rate of diamond tools is higher with silicon.

For an interrupted cutting test, the cutting tool is rapidly withdrawn from the workpiece and position z of the onset of the fracture damage zone on the uncut shoulder, relative to the tool centerline, can be measured (Fig. 2). By conducting experiments over a range of feedrates f, the parameters d_c and y_c can be obtained from an appropriate analysis of the experimental z vs f data [3,4]. Fig. 3 shows experimental values of d_c for (100) germanium crystals machined along a [011] direction as a function of the tool rake angle. d_c increases with increasing negative rake angle because of the change in stress loading conditions; larger negative rake angles suppress fracture due to the larger compressive hydrostatic component of the stress that develops with the increased tool thrust forces [5]. The results are also affected by crystallography (cutting plane and direction). Crystallography effects can be explained by assuming that microfracture during machining is generated most easily on {111} cleavage planes. This produces characteristic damage patterns on the crystal surfaces [6].

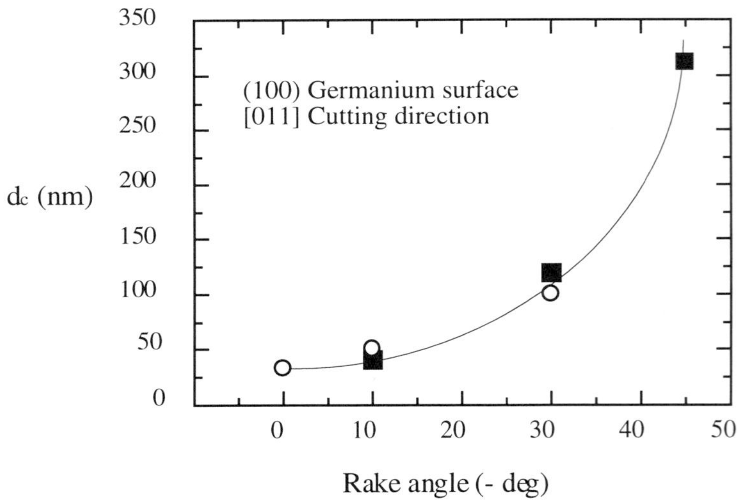

Figure 3 Measured values of d_c from interrupted cutting tests on (100) germanium crystals for a [011] cutting direction. Data from [3] (circles) and [4] (squares).

For given machining conditions and a specified material, the parameters d_c and y_c will determine the maximum feedrate for machining in the ductile regime, i.e., the value f_{max} at which the fracture damage in Fig. 2 first reaches the plane of the finished surface. For values of $f \leq f_{max}$, there will be no microfracture damage introduced into the finished surface. Analysis of the chip profile geometry in Fig. 2 gives the relationship,

$$f_{max} = d_c \sqrt{\frac{R}{d_c + 2y_c}} \qquad (1)$$

where R is the tool nose radius. The results from interrupted cutting tests have confirmed this relationship and have provided the most direct verification for the importance of chip profile geometry in controlling the fracture damage introduced into components by machining [3,4]. d_c and y_c are measured and the predicted ductile-regime limit f_{fmax} in eq. (1) can be compared to the f_{fmax} determined experimentally.

MATERIAL RESPONSE AND SCRATCH-TYPE MICROCUTTING TESTS

In order to extend the concepts of chip geometry to grinding operations, and in particular with the goal of developing process models that can be used to optimize the selection of grinding parameters for specified materials, it will be necessary to characterize parameters like d_c and y_c for grinding grains rather than large, round-nose single-point cutting tools. Interrupted cutting tests are impractical for characterization using grinding wheels since the wheel is carried on a separate spindle which cannot easily be retracted with a small PZT actuator. The characterization tests for d_c and y_c will provide an essential connection to material microstructure effects, especially at the small size scales at which fracture damage occurs for grinding grains. In higher toughness ceramics where the microstructure is designed to produce constraints that enhance toughness, for example, by means of crack bridging, the microstructural toughening effects can be counterproductive for grinding. This results from the fact that toughening mechanisms require significant crack extension in order fully develop crack-tip shielding constraints that increase toughness. A large amount of research effort has been devoted to this area and it is now widely accepted that toughness can be crack- size dependent; this is termed R-curve or T-curve behavior [7,8]. Fig. 4 shows recent toughness vs crack size data for a relatively high-toughness glass ceramic. The data were obtained using measurements of crack size under biaxial loading. Small Vickers indentation precracks were used to generate the cracks [9]. Careful analysis of the test data is needed in order to obtain objective values of the toughness using indentation precracks, as has been discussed by numerous authors [10-12].

The important implications of a crack-size dependent toughness, and especially the low toughness that can prevail for small grinding cracks in a "high toughness" ceramic, is evident from the various analyses that have been made for crack-initiation thresholds using indentation cracks. Expressed in terms of a critical depth d_c, this can be written as,

$$d_c = \beta\left[\frac{E}{H}\right]\left[\frac{K_c}{H}\right]^2 \qquad (2)$$

where K_c is fracture toughness, H is hardness, E is the elastic modulus and β is a constant depending upon indentor geometry and test conditions. This relationship was verified using acoustic emission techniques to monitor the onset of crack initiation during static indentation [13]. Bifano, et. al. [14] verified a similar relationship using a plunge-grinding geometry where the fracture initiation was determined from analysis of the ground surfaces for a range of grinding infeeds. This geometry will be considered in more detail in the next section. It should be clear that the operative toughness governing the actual d_c in a grinding operation will occur at crack size scales in the micrometer range, and this may be quite low for what is normally considered a high-toughness material. Such effects were in fact observed in the plunge-grinding studies [14]. Other manifestations of the size-scale effect for toughness will include grain-boundary cracking, residual stress variations between grains and grain pullouts during grinding.

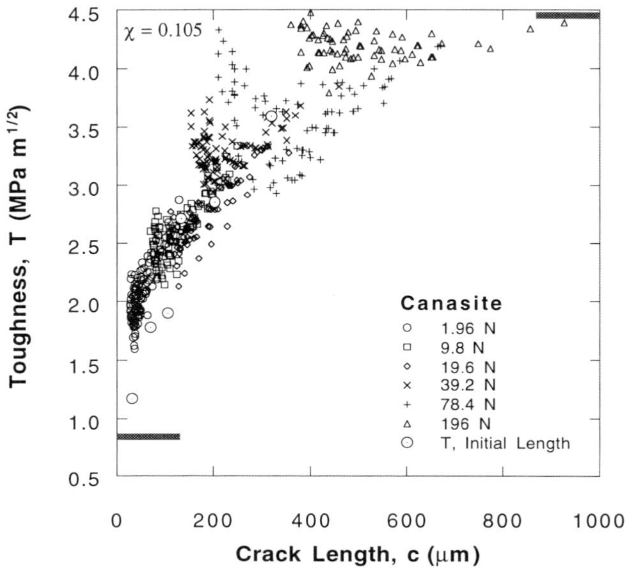

Figure 4. Fracture toughness as a function of crack length for the glass ceramic Canasite [9]. The horizontal bars indicate the toughness of the base glass (left) and the long-crack toughness measured by conventional toughness test methods (right). Crack length was varied using the range of indentation precrack loads indicated.

As was noted above, one of the critical issues for developing a process model for grinding is the determination of the parameters d_c and y_c for a "typical" grinding grain. Static indentation tests are not suitable since they do not include the cutting (tangential) force component, and factors such as the rake angle, tool nose geometry and cutting speed cannot be taken into account. Development and analysis of scratch-type microcutting tests has therefore been an important part of recent research work on the fundamentals of grinding for brittle materials. In this type of test a cutting tool, which is often a Vickers or Rockwell diamond indentor, is driven along the workpiece under a fixed load. Fracture damage occurs within the cut groove above a critical load and the general features of the fracture damage closely resemble that found for a grinding operation. Using a novel cross sectioning technique developed by Lawn and coworkers [15], Jahanimir and coworkers [16,17] used scratch-type microcutting test techniques to study the various kinds of subsurface damage produced in ceramics by a conical (Rockwell) indentor cutting tool. Comparison with actual grinding tests provided unique insight into the material removal mechanisms and confirmed the strong influence of R-curve effects in higher toughness materials. These results did not, however, provide any quantification of parameters like critical depth-of-cut d_c and subsurface damage size scale y_c.

A controlled-load microcutting test using a Vickers diamond indentor to simulate grinding grains was recently reported by Tanikella and Scattergood [18,19]. A PZT actuator loading system produces a controlled microcutting profile by ramping the indentor into the moving workpiece at a fixed loading rate, holding and then ramping out [19]. The onset of fracture at a critical load P_c was determined using acoustic emission monitoring during the microcutting test (load is related to critical depth d_c via the indentor geometry). A more direct measure of d_c can be obtained if the cutting depth is controlled rather than the load. Preliminary work [20] indicates that d_c can be determined from loading pulses at fracture initiation in a controlled-depth microcutting test if a very high stiffness load transducer is used.

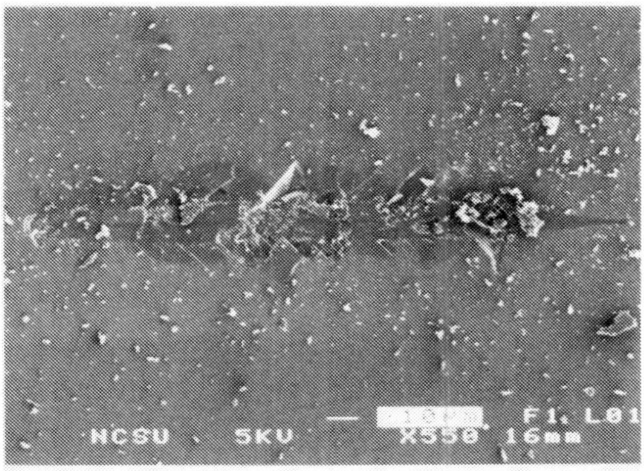

Figure 5. Microcutting groove on (100) silicon in a [011] direction made using a maximum load of 0.5 N. Cutting was parallel to the indentor side [20].

Fig. 5 shows a typical SEM micrograph of fracture damage along a microcutting groove for tests made using semiconductor grade, p-type (100) silicon surfaces cut in the [011] direction. The critical cutting depths d_c are the order of 200-400 nm; d_c varies with crystallographic surface orientation, cutting direction, loading rate and tool geometry (cutting parallel to the side or diagonal of the Vickers indentor). The groove in Fig. 5 was made with a maximum load of 0.5 N, which was above the fracture threshold at 0.10 N determined using acoustic emission monitoring. Detailed analysis showed that the subthreshold grooves at maximum loads below 0.10 N were fully ductile. Machining chips observed on the surface look very similar to those obtained in metal cutting, except that the size scale is much less. MicroRaman data indicates that the chips and the grooves contain amorphous silicon, which suggests that during the machining of silicon the metallic ß-tin structure has formed due to the high hydrostatic pressure generated by the tool loading [21]. Upon removal of the load, the material is "trapped" in the metastable amorphous state as it tries to revert to the diamond cubic silicon structure. This pressure transition phenomenon is a unique feature for the machining of certain brittle materials such as silicon or germanium. It has been observed in past research using static indentation [21] and other scratch-type microcutting tests [22].

The fracture damage morphology and maximum subsurface fracture damage depth y_c also vary with crystallographic cutting direction, cutting geometry and loading rate for a test like that shown in Fig. 5. The y_c values are larger than the d_c values and in unfavorable cases the differences can be an order of magnitude or more [19]. The characterization of the fracture damage morphology and the determination of y_c parameters requires a suitable technique for preparing cross sections of the microcutting grooves. This can be difficult for materials than cannot be cleaved or broken to produce smooth fracture surfaces. Further development work is necessary for obtaining the required cross sections on microstructurally complex materials such as polycrystalline ceramics. The technique developed by Lawn and coworkers [15], using a split/clamped sample with prepolished cross sections, is one possibility although sample preparation difficulties could be a concern.

PLUNGE GRINDING

A central issue in the development of a process model for grinding is the relationship between the influence of individual cutting grains, as manifest by parameters like d_c and y_c, and an effective cutting depth based on the uncut chip profile. This relationship was easily visualized for single-point machining because the uncut chip profile in Fig. 2 directly reflects the geometry of the cutting tool edge and the chip removal is continuous. With a grinding operation, the uncut chip profile is determined by the profile of the grinding wheel and the crossfeed and upfeed motions for the particular type of grinding operation (contour, cup wheel, etc.). Furthermore, the wheel rotates and contains within the binder a distribution of grinding grains that produce a multi-point cutting action. The average grain depth-of-cut d_g determines when individual grains cause a ductile-to-brittle transition along the chip profile boundary. As a more graphic illustration, Fig. 6 shows an uncut chip profile for single-point machining (Fig. 2) and a similar chip profile with the tool edge replaced by an embedded distribution of cutting grains. The effective cutting depth d used in single-point machining must be replaced by an effective depth-of-cut per grain d_g. As will be discussed further, the latter will depend upon the number of active cutting grains that participate in material removal. This will be determined by the grinding wheel rotation and associated wheel runout as well as the distribution of the grains in the binder and related factors.

The primary effects can be visualized using plunge-grinding geometry. In this case, there is no crossfeed or upfeed motion of the grinding wheel relative to the workpiece; the wheel is simply plunged vertically into a stationary workpiece at a specified infeed rate. A more difficult and crucial issue is the determination of the effective depth-of-cut per revolution of the wheel, i.e., the uncut chip profile as determined from the grinding wheel profile without regard for the grinding grains. If the wheel ran perfectly true about its rotation axis, the cutting action would be continuous and a finite depth-of-cut would be not be defined. However, a grinding wheel can never be made perfectly true and there will also be synchronous error motion due to machine dynamics. Consideration of these errors, in conjunction with the submicron d_c values to be expected for typical brittle materials, implies that the grinding wheel will act as a flycutter. There will be a once-per-revolution cutting action caused by the "high spot" on the grinding wheel. The effective depth-of-cut per revolution based on the wheel profile will therefore be the infeed rate f used for the plunge grinding.

The measurement of grinding force during plunge grinding gives a direct means of verifying a once-per-revolution grinding action. Fig. 7 shows the normal (thrust) force measured using a very stiff high-precision plunge-grinding machine [23]. Even with a very carefully trued and dressed grinding wheel, force peaks occur with a period equal to the rotational period of the wheel. This is a clear signature of a once-per-revolution flycutting action, or "high-spot" mode for the grinding operation. Furthermore, the peak-to-valley variation in the grinding force will increase if the runout errors are increased, as would be expected for high-spot grinding [23]. This effect has also been verified for other modes of grinding using crossfeed and upfeed motions. The once-per-revolution cutting action that ultimately defines the chip-profile geometry produces a pattern on the workpiece surface that has a characteristic phasing shift caused by successive passes of the high spot as the workpiece moves relative to the grinding wheel [24,25].

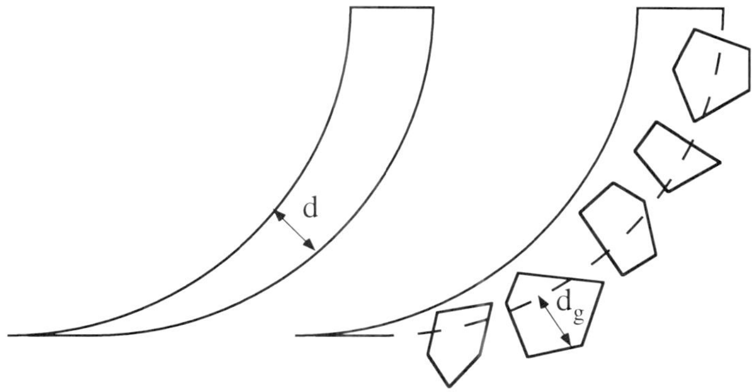

Figure 6. Uncut chip profile for single-point machining (left) and a similar profile with cutting grains present (right). When grains act as the cutters, the grain depth-of-cut d_g replaces the single-point tool depth-of-cut d.

In order to develop a physically realistic process model for grinding, it will be necessary to relate the critical depth parameters d_c and y_c for single cutting grains, as discussed in the previous section, with the operational limits for the particular grinding operation. The limiting feedrates required to remain within the ductile grinding regime are of particular importance in this regard since they provide a baseline for subsurface damage generation. Simulation results recently developed for plunge grinding show the significant aspects of this kind of relationship [26]. The chip profile for plunge grinding, assuming the flycutting mode, is the cross section fW shown in Fig. 8 where f is the infeed per revolution of the wheel and W is the wheel width. This is the uncut chip profile as determined by the grinding wheel profile. However, the cutting grains must be overlaid onto the chip profile as was indicated in Fig. 6. The critical depth-of-cut for a ductile-to-brittle transition will then be determined by the mean cutting depth per grain d_g, or equivalently the cutting area per grain A_g, as shown in Fig. 8. The area A_g can be determined from the steady-state requirement $N_cA_g = fW$ shown in Fig. 8, viz., the cross section removed per grinding-wheel revolution fW must be equal to the projected cutting cross section per grain A_g times the number of active cutting grains N_c on the wheel high spot.

Simulations have been made for plunge grinding which take into account features such as grain distribution, grain shape, binder compliance and grain shadowing [26]. Shadowing is the reduced cutting action of grains immediately following the lead grains on the high spot. This is determined by the mean free path between the cutting grains. If ρ is the density of the cutting grains per unit area of the workpiece cut surface at a particular value of A_g, then the mean free path L is determined by the relation $D_g\rho L = 1$ where D_g is the apparent size of the grains measured at the cut surface, not the mean size of the grains in the binder. A simulation proceeds by introducing a distribution function for the grain heights using a specified grain shape and then determining the position of the cut surface relative to the cutting grain distribution for a given value of fW. The analysis is somewhat cumbersome but straightforward provided that there is no correlation between the ground surface topography and the profile of the active cutting grains on the wheel periphery. If such a correlation exists, then the surface position shown dashed in Fig. 8 will tend to "follow" the outline of the cutting grains and the operative grain depth-of cut d_g will be significantly reduced. This does not seem likely for real grinding operations with crossfeed and upfeed since the overlap of successive cutting passes will be out of phase with previous passes at a given position on the workpiece surface.

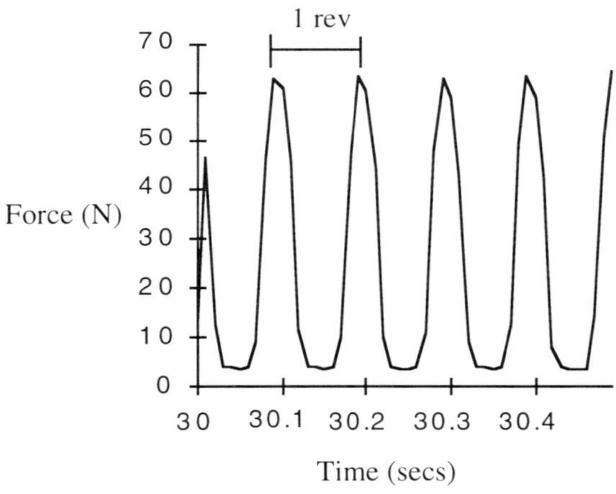

Figure 7. Normal force variations for plunge grinding of sapphire. Wheel runout was 12 μm using a mechanically dressed grinding wheel with 50 μm grain size in a metallic binder. The period for one revolution of the wheel is indicated [23].

The simulation results for plunge grinding show a number of important features. Fig. 9 shows the critical feedrate f_c for a ductile-to-brittle transition in the plunge-ground finished surface plotted vs the critical depth-of-cut d_c for an individual grain. D_{eff} is the mean size of the grains in the binder and F_vL_o/D_{eff} is a normalized parameter for the high-spot duration (fraction of the wheel circumference that contains the active cutting grains). Note that f_c is significantly less than d_c. This results from the fact that the grain depth-of-cut d_g will be much larger than the effective depth-of-cut f determined by the chip profile geometry. If there is a strong correlation between features on the ground surface and the profile of the active cutting grains on the grinding wheel periphery, then this difference will be reduced and f_c will increase. When the high-spot duration is increased at larger values of the parameter F_vL_o/D_{eff}, f_c increases because the grain depth-of-cut d_g is decreased for a given feedrate, i.e., more grains can become active cutting grains with a longer high-spot duration. Optimum values of f_c are obtained for continuous cutting (no high spot). There is an intrinsic grain size dependence, not readily evident from the grain-size normalized coordinates used in Fig. 9, such that smaller values of grain size will increase f_c for a fixed volume fraction of grinding wheel abrasive grains. This effect is observed in grinding practice, i.e., smaller grain sizes reduce the level of subsurface fracture damage for a given choice of grinding geometry and operating conditions.

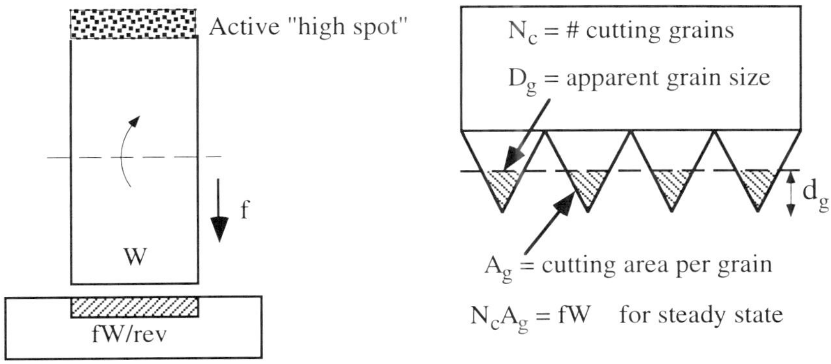

Figure 8. Plunge grinding geometry and grain depth-of-cut requirements for steady state material removal. D_g is the apparent grain size and ρ is the number density measured on a plane section at the workpiece cut surface (dashed line).

There is an interesting and unexpected "cutoff" regime where the lines in Fig. 9 change slope and become relatively independent of the high-spot duration parameter. This regime occurs when the high spot duration and the mean-free path length between active cutting grains have comparable values. Changes in d_c have a diminished influence on the limiting feedrate f_c in this regime. The effect is due to a change in the grain shadowing behavior when the high-spot duration and mean-free path length L have comparable values.

Results of the kind shown in Fig. 9 have been used to experimentally verify the relation for d_c given in eq. (2) for grinding. A selection of different materials with varying K_c/H values were plunge ground using fixed grinding conditions and the amount of fracture damage in the finished surface was determined as a function of the infeed [14]. Since there are no crossfeed or upfeed motions in plunge grinding, the ground surface will show ductile-to-brittle fracture transitions as soon as the grain depth-of-cut d_g equals d_c (compare to Fig. 2). It should be noted that the methodology used in [14] did not account for grain depth-of-cut effects, and the authors assumed that that $f_c = d_c$. For the fixed grinding conditions used, this will give suitable correlations for verifying eq. (2), but the absolute value of d_c obtained will be an underestimate of the true d_c value for a grinding grain. Using plunge grinding as a means to characterize d_c values is therefore an indirect procedure since it depends on factors such as the high-spot duration, the grain distribution and other parameters. In that context, the binder used for the grinding grains can have a strong influence on the results.

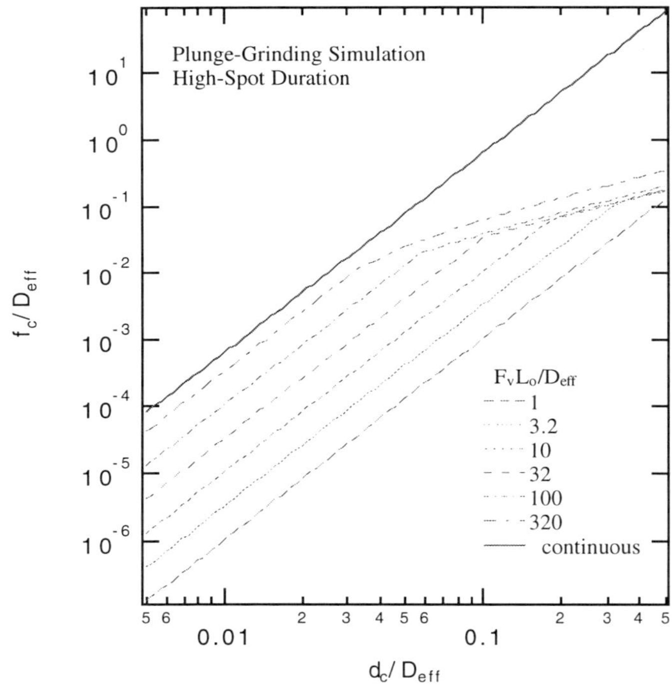

Figure 9. Simulation model showing the effect of wheel high-spot duration. A random distribution of grain heights for wedge-shaped grains was assumed and no surface correlation effects were included. Continuous cutting means the high-spot duration is the entire wheel circumference.

Fig. 10 shows simulation results similar to Fig. 9 where the binder compliance for the grains is varied by means of a normalized compliance parameter (large values of the parameter correspond to high binder compliance, or low binder stiffness). As the compliance is increased, for example, by changing from metal to resin binders, f_c increases with other conditions held fixed. This effect of softer binders, which is well known as a means of reducing subsurface damage in grinding practice, is a result of the fact that soft binders "indent", bringing more active cutting grains into contact with the workpiece. The increase in N_c reduces the mean cutting area A_g and grain depth-of-cut d_g in Fig. 8, thereby increasing the limiting infeed f_c. The disadvantage of soft binders is that there is a loss in the form accuracy that can be maintained because of the binder deflection. This can be very detrimental for applications where high precision parts are being manufactured using grinding operations, hence, metal binders are very often used.

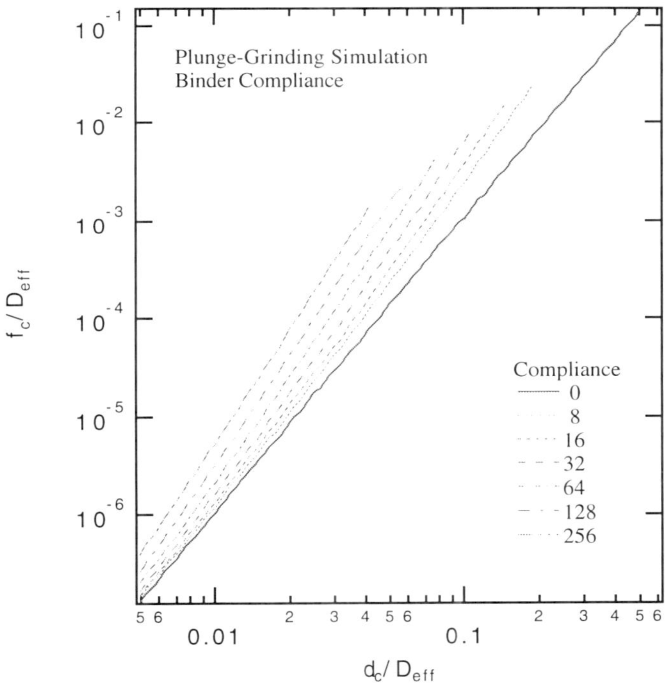

Figure 10. Simulation model showing the effect of binder compliance for continuous cutting. The simulation parameters are the same as in figure 9.

The simulation results shown in Figs. 9 and 10 are facilitated by the choice of plunge-grinding geometry. In real grinding operations using crossfeed and upfeed, the uncut chip profile geometry will be three-dimensional and very complex. Computer simulation and visualization techniques are being developed to deal with these geometrical issues [24,25]. The basic concepts remain unchanged, however, and the steady-state material removal requirement shown in Fig. 8 remains the same, except that this must be applied over all cross sections being removed along the chip profile boundaries. Depending upon the choice of operating parameters (wheel speed, part speed, etc.) there will always be a critical cross section that governs the ductile-regime limit and hence the critical feedrates for achieving ductile-regime grinding. Features such as the wheel high-spot duration and binder stiffness can be expected to play a similar role to those demonstrated here using the plunge grinding geometry.

SUMMARY

Material response characterized by fundamental damage parameters embodied in a critical depth-of-cut d_c and subsurface damage size scale y_c at the ductile-to-brittle transition, in conjunction with the uncut chip profile geometry and steady-state material removal requirements, can be used to analyze the generation of subsurface fracture damage in the machining and grinding of brittle materials. Of particular interest are the operational parameters for ductile-regime grinding, especially the critical feedrates which define baseline limits for the ductile regime. The approach outlined in this review provides a framework for the development of physically based process models for machining and grinding. Significant challenges remain in the futher development of simulation and visualization techniques, characterization of material response via critical depth parameters d_c and y_c for grinding grains and the quantification of grinding wheel parameters such as runout, high-spot duration, grain distribution and grain compliance.

Support provided by the U. S. Department of Energy under contract No. DE-FG05-84ER45115, and the Precision Engineering Center, North Carolina State University, is gratefully acknowledged.

REFERENCES

1. G. Boothroyd, “Fundamentals of Metal Machining and Machine Tools”, McGraw-Hill (1975).
2. B. Lawn, “Fracture of Brittle Solids”, 2nd edition, Cambridge Solid State Science Series, Cambridge Press (1993).
3. W. S. Blackley and R. O. Scattergood, “Ductile-Regime Machining Model for Diamond Turning of Brittle Materials”, Precis. Engr., **13** [2] 95 (1991).
4. R. M. Tidwell, “Ductle Regime Machinine of Germanium: Development of New Experimental and Analytical Analysis Methods”, MS Thesis, North Carolina State University (1991).
5. K. Ueda, T. Sugita and H. Tsuwa, “Application of Fracture Mechanics in Micro-Cutting of Engineering Ceramics”, Annals of the CIRP, **32** [1] 83-86 (1983).
6. W. S. Blackley and R. O. Scattergood, "Crystal Orientation Dependence of Machining Damage - A Stress Model", J. Am. Ceram. Soc., **73** [10] 3113 (1990).
7. R. F. Cook, B. R. Lawn and C. J. Fairbanks, “Microstructure-Strength Properties in Ceramics”, J. Am. Ceram. Soc., **68** [11] 604-15 (1985).
8. S. J. Bennison and B. R. Lawn, “Role of Interfacial Grain-Bridging Sliding Friction in the Crack-Resistance and Strength Properties of Non-Transforming Ceramics”, Acta Metall., **37** [10], 2659-72 (1989).
9. S. M. Smith and R. O. Scattergood, “Determination of Short-Crack Toughness Curves”, J. Am. Ceram. Soc., in press (1995).
10. L. M. Braun, S. J. Bennison and B. R. Lawn, “Objective Evaluation of Short-Crack Toughness Curves Using Indentation Flaws”, J. Am. Ceram. Soc., **75** [11] 3049-57 (1992).

11. R. F. Krause, "Flat and Rising R-Curves for Elliptical Surface Cracks from Indentation and Superposed Flexure", J. Am. Ceram. Soc., **77** [1], 172-78 (1994).
12. Bleise and R. F. Steinbrech, "Flat R-Curve from Stable Propagation of Indentation Cracks in Coarse-Grained Alumina", J. Am. Ceram. Soc., **77** [2] 315-22 (1994).
13. J. Lankford and D. L. Davidson, "The Crack Initiation Threshold in Ceramic Materials Subject to Elastic/Plastic Indentation", J. Mater. Sci., **12** 1662-68 (1979).
14. T. G. Bifano, T. A. Dow and R. O. Scattergood, "Ductile-Regime Grinding: A New Technology for Machining Brittle Materials", Jour. of Engr. for Industry, **113** 184-89 (1991).
15. F. Guiberteau, N. P. Padture and B. R. Lawn, "Effect of Grain Size on Hertzian Contact Damage in Alumina", J. Am. Ceram. Soc., **77** [7] 1825-31 (1994)
16. H. H. K. Xu, S. Jahanimir and Y. Wang, "Effect of Grain Size on Scratch Interactions and Material Removal in Alumina", J. Am. Ceram. Soc., **78** [4] 881-91 (1995).
17. H. H. K. Xu, N. P. Padture and S. Jahanimir, "Effect of Microstructure on Material-Removal Mechanisms and Damage Tolerance in Abrasive Machining of Silicon Carbide", j. Am. Ceram. Soc., **78** [9] 2443-48 (1995).
18. B. V. Tanikella, K. A. Gruss, R. F. Davis and R. O. Scattergood, "Indentation and Microcutting Fracture Damage in a Silicon Carbide Coating on an Incoloy Substrate", submitted to Surf. and Coating Tech. (1995).
19. B. V. Tanikella and R. O. Scattergood, "Fracture Damage in Borosilicate Glass during Microcutting Tests", Scripta Met. et. Mater., in press (1995).
20. B. V. Tanikell, PhD thesis dissertation in progress, North Carolina State University (1995).
21. D. R. Clarke, M. C. Kroll, P. D. Kirchner, R. F. Cook and B. J. Hockey, "Amorphitization and Conductivity of Silicon and Germanium During Indentation", Phys. Rev. Lett., **60** 2156 (1988).
22. J. C. Morris, D. L. Callahan, J. Kulik, J. A. Patten and R. O. Scattergood, "Origins of the Ductile Regime in Single-Point Diamond Turning of Semiconductors", J. Am. Ceram. Soc., **78** [8] 2015 (1995).
23. K. W. Sharp and R. O. Scattergood, "In-Process Dressing of Grinding Wheels for Brittle Materials", Proc. ASPE Annual Mtg., Austin, Texas, 195 (1995).
24. S. C. Fawcett and T. A. Dow, "Influence of Wheel Speed on Surface Finish and Chip Geometry in Precision Contour Grinding", Precis. Engr., **14** [3] 160-67 (1992).
25. G. E. Storz and T. A. Dow, "Cup Wheel Grinding Geometry", Proc. ASPE Annual Mtg., Cincinnati, Ohio, 105 (1994).
26. R. O. Scattergood, to be submitted for publication (1995).

USE OF BOUNDARY ELEMENT METHOD TO DESIGN GLASS AND METAL COMPONENTS

Asish Ghosh and Chin Young Cha
Philips Display Components Company, Ann Arbor, U.S.A.

ABSTRACT

Finite element method has been used extensively to design metal components. However, the use of computational methods such as finite element methods to design glass and ceramic components is still being developed. Boundary element method is another computational technique which has grown considerably over the last decade. Boundary element method is evaluated in this study to see if it can be used for designing glass and metal components. The advantages and limitations will also be discussed.

INTRODUCTION

Computational techniques such as finite element method (FEM) has been used to design glass and ceramic components[1,2]. It has also been used, in conjunction with statistical reliability analysis techniques[3], to predict probability of failures for the components[4].

Limited amount of work has been done to use Boundary Element Method (BEM) to design glass and metal components[5-8]. However, to our knowledge even less work has been done to use BEM to determine manufacturability of these components.

In this paper a review of the use of BEM to design cathode ray tube (CRT) components is made. Examples of a 27 inch diagonal and a 32 inch diagonal CRT are considered. Deflection and stresses due to evacuation of glass envelopes as well as due to the application of metal shrinkband around the panel glass of evacuated glass envelopes are determined. Stresses generated due to metal clamps, around the neck glass of evacuated glass envelopes, to attach deflection yokes are also examined in this work.

DEFLECTION AND STRESS ANALYSIS DUE TO EVACUATION AND SHRINKBAND APPLICATION

The case considered here is to determine the maximum principal stresses during evacuation of the CRT. This process step is called vacuum case. In order to provide adequate safety to the CRT, in the event of an implosion, a metal band is applied around the circumference of the panel glass (Refer Figure 1). This generates compressive stresses in the panel and thereby help reduce the tensile stresses generated during evacuation. This process step is called shrinkband case.

During evacuation the center of the front face displaces towards the funnel. However, upon application of the metal shrinkband, the center face tries to go back to its original position. The amount of displacement, by which the center face moves away from the funnel upon application of the shrinkband, is called "doming". Normally the doming value is anywhere between 40% and 60% of the amount of displacement by which the center face moves during evacuation.

Comparison of BEM and FEM

Figure 1 exhibits the boundary element model (Quarter) of a CRT. The BEM model consisted entirely of Q38 type, quadratic quadrilateral boundary elements. The FEM model consisted of a mixture of STIF 95, 20 node isoparametric solid elements and STIF 93 eight node isoparametric shell elements. Figure 2 and 3 exhibits the maximum principal stresses in the vacuum case for BEM and FEM. In both cases the maximum tensile stresses due to evacuation is on the outside surface, at the middle of the long side, of the panel face. The maximum stresses considering one element through the thickness are 1452.6 psi using BEM and 1306 psi using FEM. This is a difference of about 10% between the two methods.

A convergence study of BEM and FEM indicated that [7] the total number of elements and modeling time are less for BEM compared to FEM. Table I tabulates the timings and the number of elements required through the thickness to converge. As the number of elements increase from one to two, the time taken by FEM to converge, compared to BEM, increases by almost two

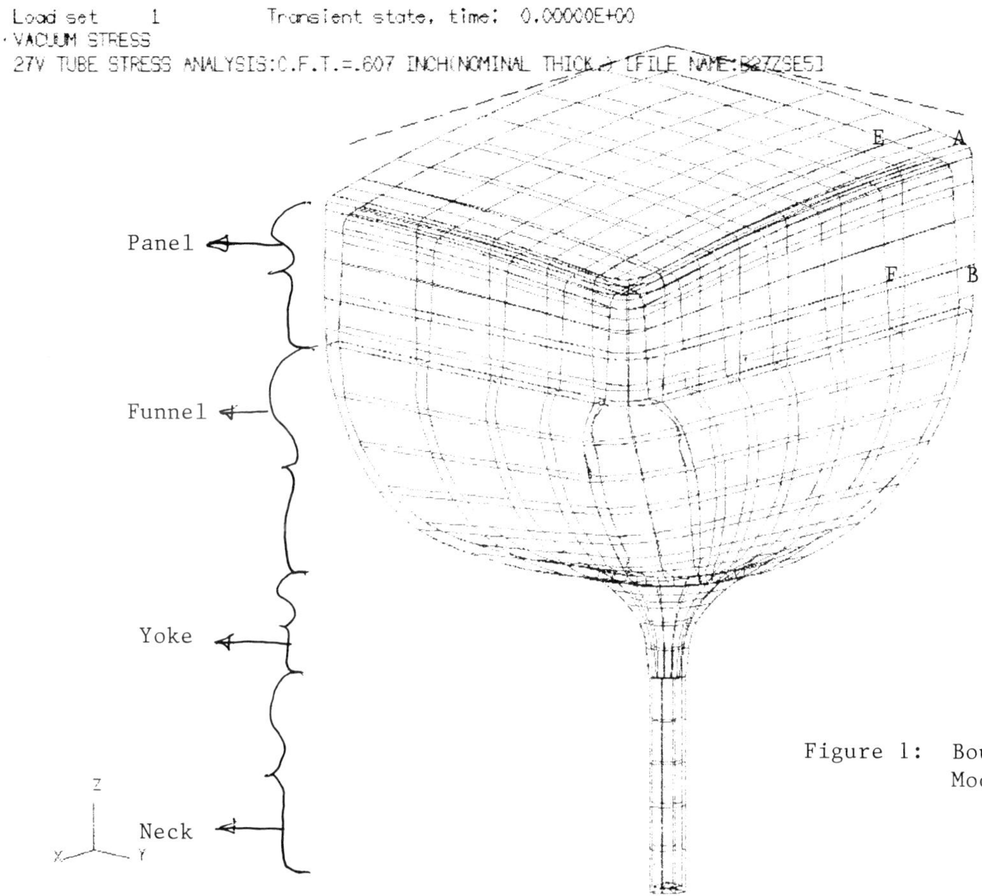

Figure 1: Boundary Element Model of a 27V CRT

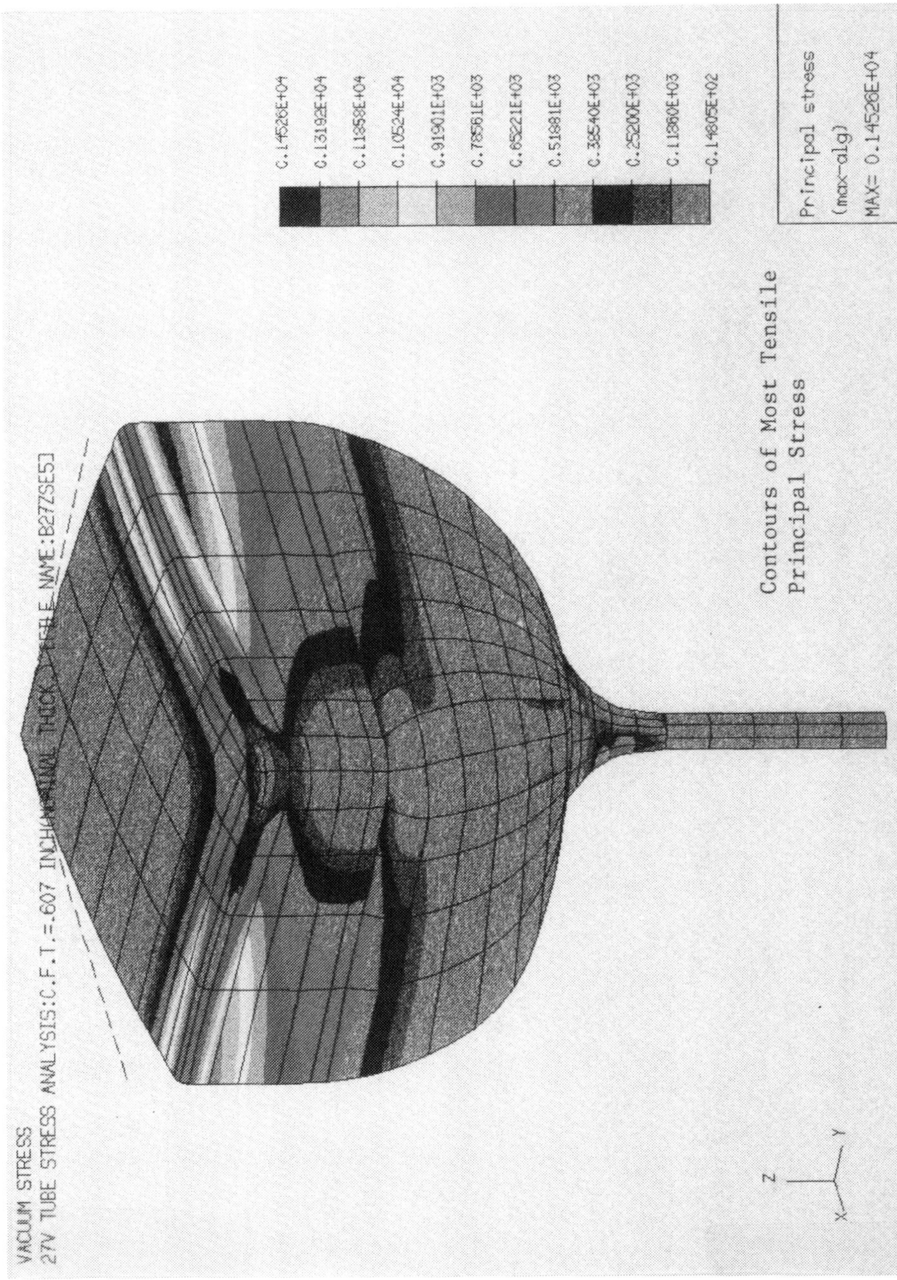

Figure 2: Boundary Element Stress Results of a 27V CRT.

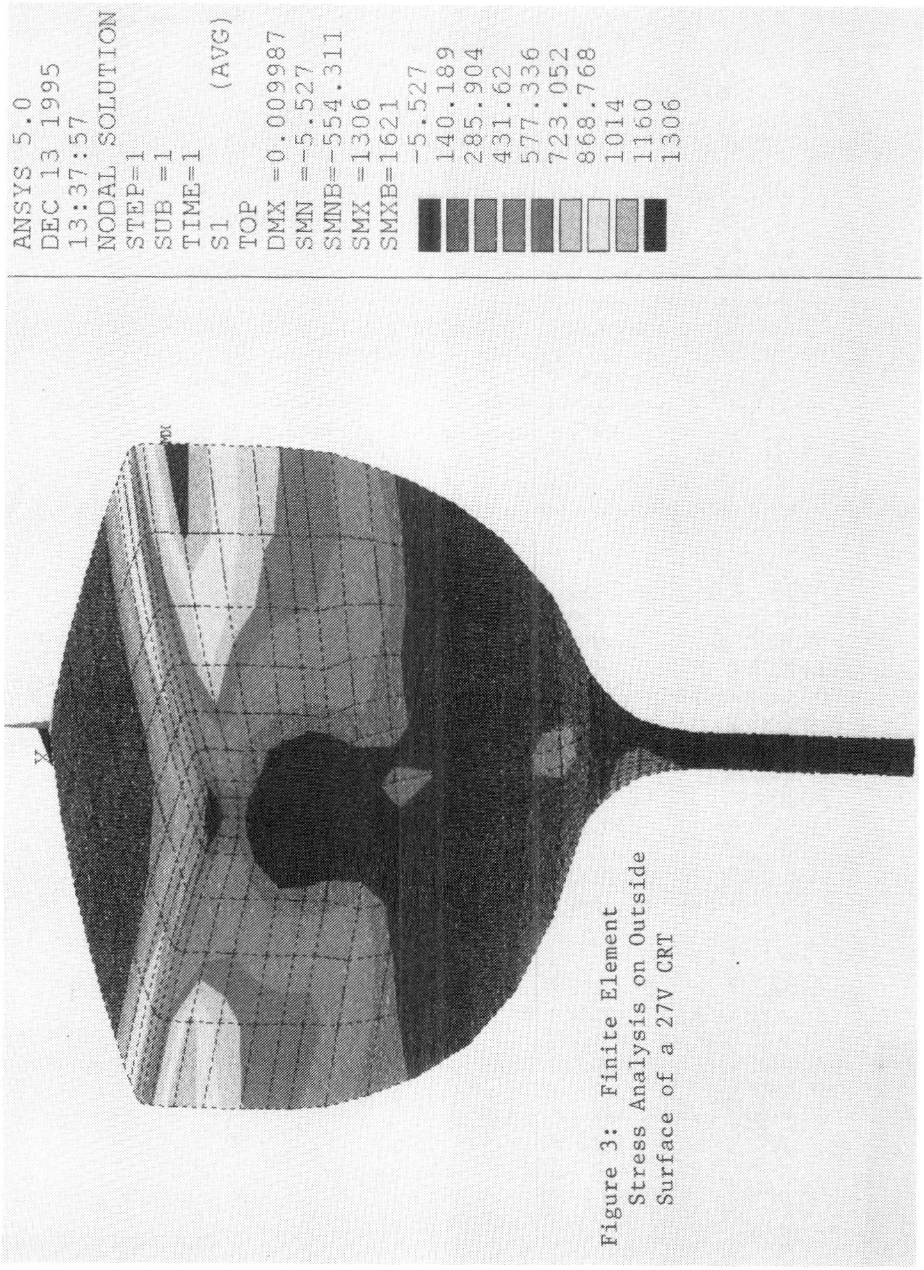

Figure 3: Finite Element Stress Analysis on Outside Surface of a 27V CRT

TABLE I: SUMMARY OF MODELING TIMES FOR FEM AND BEM

NUMBER OF ELEMENTS/ THICKNESS	ANALYSIS TYPE	TOTAL NUMBER OF ELEMENTS	MODELING TIME	RATIO FEM/ BEM
1	BEM	542	6 HOURS	2.7
	FEM	1152	16 HOURS	
2	BEM	566	5 MINUTES	96
	FEM	1536	8 HOURS	
3	BEM	590	5 MINUTES	96
	FEM	1920	8 HOURS	

TABLE II: MAXIMUM PRINCIPAL STRESSES USING BEM

NUMBER OF ELEMENTS/ THICKNESS	LOADING CONDITION	MAX. PRINCIPAL STRESS(PSI)	DEVIATION
1	VACUUM	1452.6	0.14%
	SHRINKBAND	1195	0.084%
2	VACUUM	1451	0.028%
	SHRINKBAND	1194	0%
3	VACUUM	1450.6	0%
	SHRINKBAND	1194	0%

TABLE III: MAXIMUM PRINCIPAL STRESSES USING FEM

NUMBER OF ELEMENTS/ THICKNESS	LOADING CONDITION	MAX. PRINCIPAL STRESS(PSI)	DEVIATION
1	VACUUM	1306	11%
	SHRINKBAND	1059	11%
2	VACUUM	1458	0.81%
	SHRINKBAND	1186	0.75%
3	VACUUM	1470	0%
	SHRINKBAND	1195	0%

orders of magnitude (2.7 to 96).

Table II and Table III summarize the maximum principal stresses using BEM and FEM. "Deviation" is the percentage error assuming the exact solution is as given by the analytical model with three elements in the thickness direction, which is the most refined model. The deviation in stresses are less than 0.5% for one element through the thickness for BEM, while it takes three elements through the thickness for FEM to get to that low deviation values. The deviation in vacuum stresses is 11% for FEM for one element through the thickness.

Upon application of the shrinkband, the stresses are lower than in the vacuum case. For BEM it is 1195 psi compared to 1059 psi for FEM using one element through the thickness. Similar trends in deviation are observed in the shrinkband case for different number of elements through the thickness.

For design and to determine manufacturability as well as to predict long term reliability, the stresses that need to be considered are the permanent stresses. This is the shrinkband case. In this example, therefore, the permanent stresses in a 27V CRT is 1195 psi, which is the ideal case for both BEM (one element) and FEM (three elements). Thus BEM is a faster method to determine manufacturability.

Table IV lists the "doming" values as described earlier. In this case significant difference is not observed between FEM (0.089 mm) and BEM (0.087 mm). This is expected because doming is a measure of displacement and FEM is considered to be as good as BEM for predicting displacements. Stress values, however, involves taking derivatives and this is where significant differences in deviation are observed between BEM and FEM.

Figure 4 exhibits the stresses (mostly tensile) along section AB on the outside surface of the tube (Refer Figure 1). Figure 5 exhibits the stresses (mostly compressive) along section EF, which is the inside surface of the tube. The stress profiles are considered at this location because this is the region of high stresses as shown in Figure 2. Since the outside surface of glass has a higher probability of having flaws compared to the inside surface therefore,

TABLE IV: BOUNDARY ELEMENT AND FINITE ELEMENT RESULTS FOR DOMING

NUMBER OF ELEMENTS/ THICKNESS	DOMING VALUE FROM BEM	DOMING VALUE FROM FEM
1	0.087 MM	0.088 MM
2	0.087 MM	0.088 MM
3	0.087 MM	0.089 MM

TABLE V: COMPARISON OF STRESSES FOR THREE DESIGNS AT POINT LOCATIONS A, B AND C

DESIGN	VARIOUS STRESS	POINT A	POINT B	POINT C
PRESENT DESIGN	YOKE CLAMP STRESS (PSI)	2993	3040	7942.8
	YOKE RESIDUAL STRESS (PSI)	609.4	13.74	0
	YOKE TOTAL STRESS (PSI)	3448	3034	7687.8
OPTION I	YOKE CLAMP STRESS (PSI)	2450	2489	6501.6
	YOKE TOTAL STRESS (PSI)	2905	2482	6246.6
OPTION II	YOKE CLAMP STRESS (PSI)	2602	2465	4767.4
	YOKE TOTAL STRESS (PSI)	3032	2454	4595.4

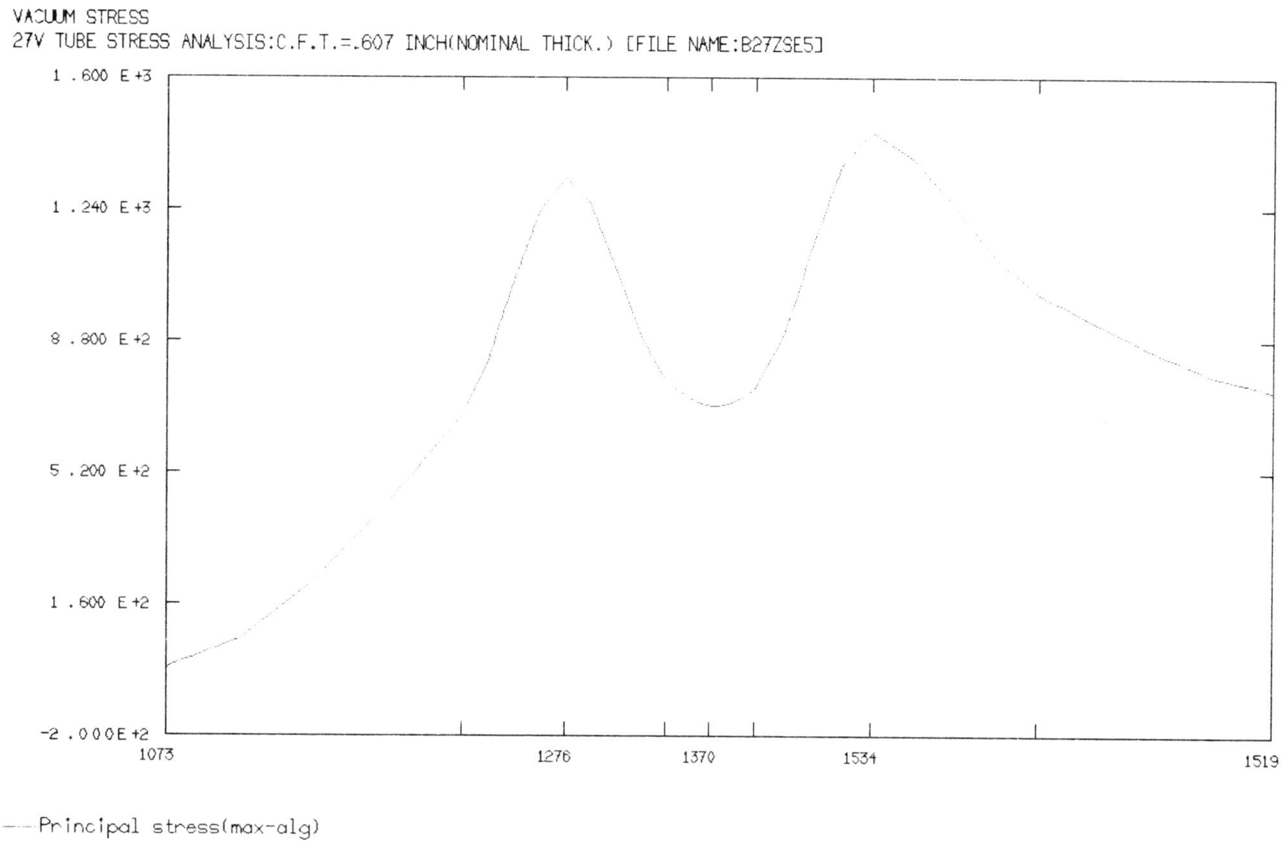

Figure 4: Boundary Element Results. Tensile Stress Along A-B

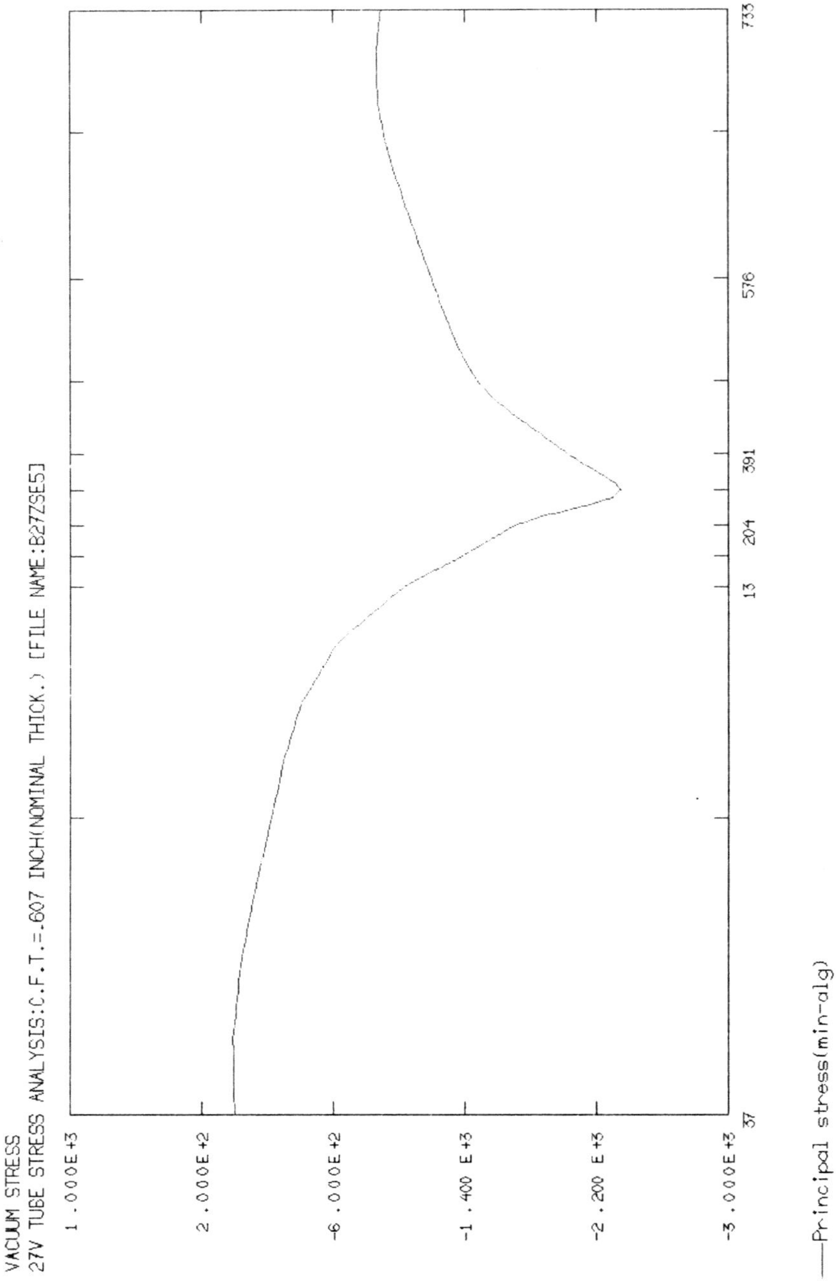

Figure 5: Boundary Element Results Compression Mesh point averages
Stress Along E-F

it can fracture at lower tensile stresses[1,4]. It is critical that the magnitude of the tensile stress on the outside surface should not exceed the strength of glass[3]. The strength of glass is determined by considering the flaw size distribution on the outside surface[2-4]. In this case, the maximum tensile stress was less than the strength of glass. Therefore the design was considered acceptable.

The ability to profile the stresses along the thickness and along sections of both inside and outside surfaces is a valuable feature of BEM. This is used to determine the optimal manufacturable design. The stress profile can be analyzed to determine the regions of high stresses and then determine ways to reduce the high stresses in these regions. In this case the design was manufacturable, therefore no change was made to the design.

STRESS ANALYSIS OF NECK CLAMPS

To determine another aspect of manufacturability of CRT, the stress analysis for applying yokes around the neck of a CRT (Refer Figure 1) needs to be done. The yokes provide the electromagnetic field to the electron beam for the desired sweeping on the front face. Figure 6 provides an enlarged picture of the neck region along with the locations of point A,B and C. These are the points where the stresses are determined.

Table V and VI summarize the stresses at the various locations for two different options and the present situation. Total stresses in the neck glass are a combination of the neck clamp stresses as well as the residual stresses in the glass. The residual stresses in the glass comes about from the processing history of the neck glass.

As given in Table V the stresses are the lowest at point C for Option II (4595.4 psi) compared to the present design (7687.8 psi) and Option I (6246.6 psi). However, actual analysis of crack pattern in fractured necks indicate that the critical location is point A. The stresses at point A are lower for Option I (2905 psi) compared to present design (3448 psi) and comparable to Option II (3032 psi). At point B, trends similar to Point A are observed. From the above the observation is that, either of the Options I or II is

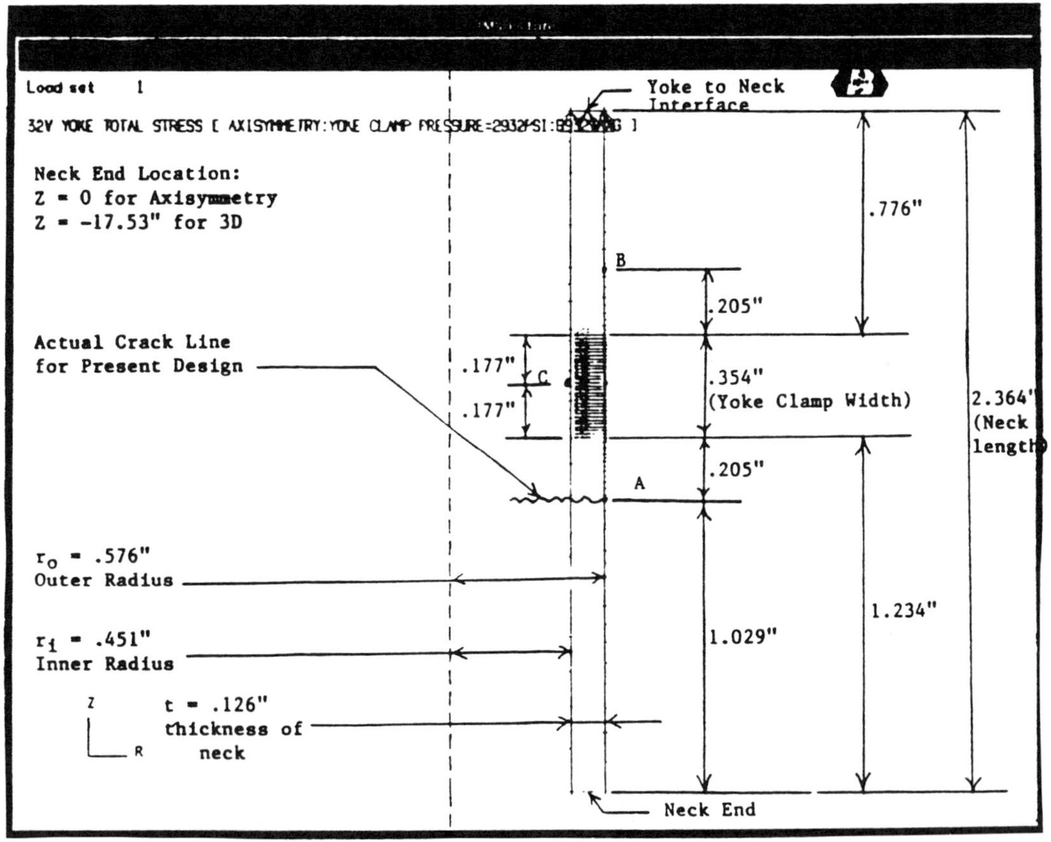

Figure 6: Yoke Clamp Location for Present Design & Option 1 with Boundary Element Model with 32V Neck

TABLE VI: SUMMARY OF VARIOUS RESULTS FOR DIFFERENT NECK CLAMP DESIGN OPTIONS

ITEMS	PRESENT DESIGN	DESIGN OPTION I	DESIGN OPTION II
YOKE CLAMP TORQUE	1 N-m (8.85 lbf-in)	0.8181 N-m (7.24 lbf-in)	1 N-m (8.85 lbf-in)
WIDTH OF YOKE CLAMP	9 MM (0.354")	9 MM (0.354")	12.6 MM (0.498")
YOKE TOTAL STRESS (A)	3448 PSI	2905 PSI	3032 PSI
NORMAL FORCE TO THE NECK	3756 lb	3074 lb	3756 lb
YOKE CLAMP PRESSURE	2932 PSI	2400 PSI	2084 PSI
CONDITION	PRESENT CONDITION	REDUCE YOKE CLAMP TORQUE TO 0.818 N-m ON PRESENT CONDITION	NEW DESIGN WITH HIGHER YOKE CLAMP WIDTH OF 12.6 MM
RESULT	STRESS FAILURE COULD OCCUR AT POINT A	NO STRESS FAILURE EXPECTED AT POINT A	NO STRESS FAILURE EXPECTED AT POINT A

acceptable.

Table VI lists the various items for the different options and the present design. It is clear that going to Option II in order to reduce stresses at point C as opposed to staying with the present design will require different clamp width (12.6 mm vs. 9 mm) i.e. a different design of clamp. This will result in increase in cost and possible changes in the application tool. However, Option I just required reducing the neck clamp torque from 1 N-m for the present design to 0.8 N-m, and kept the clamp design same as the present design. This reduced the stresses in location A but not in location C. Since location A is the most critical and stresses could be reduced from 3448 psi to 2905 psi just by reducing the torque, this was a manufacturable option (less cost and less changes compared to the present condition) and was chosen as a process step.

Figure 7 exhibits the profile of total stresses in the yoke region for the present design. This is the advantage of using BEM i.e. the stress profiles can be plotted and compared for the various options to determine the optimal process step.

From the above it is clear that by using BEM, optimal design and manufacturable steps can be determined theoretically. The option of plotting stress profiles is also extremely helpful.

CONCLUSIONS

BEM converges faster than FEM by about two orders of magnitude when more than two elements are used through the thickness. The deviation in stress values is less than 0.5% for BEM for one element through the thickness. For predicting displacement values BEM and FEM are comparable. However, BEM can be used to determine the optimal manufacturable CRT envelope design (glass) and neck clamp design (metal) at faster rates, because stresses are also considered to determine manufacturability.

REFERENCES

1)A. Ghosh, C.Y. Cha, S. Vaidyanathan and R.C. Bradt, "Finite Element Stress Analysis and Crack Path Prediction of Imploding CRT," Ceramic Transactions,

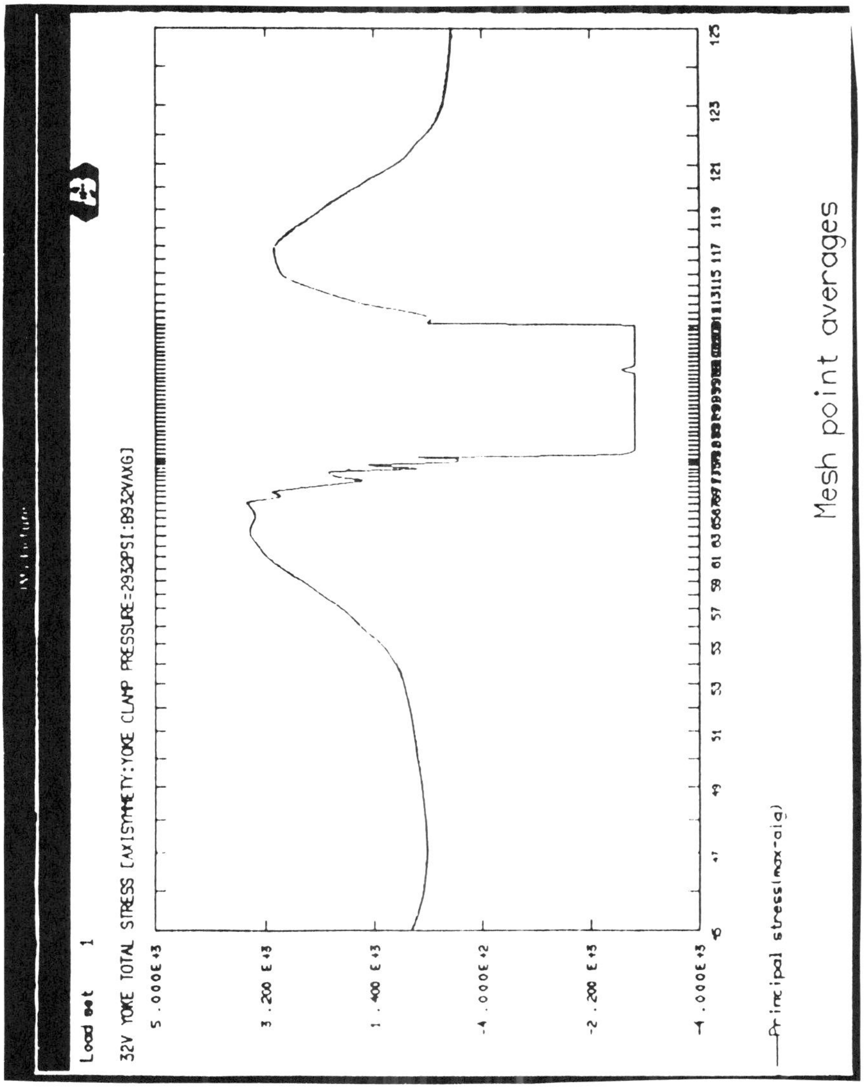

Figure 7: Graph of Total Stress in Yoke Region (Current Design)

Volume 17.

2)A. Ghosh, C.Y. Cha and S. Vaidyanathan, "Design for Manufacturability of Cathode Ray Tubes," pp 348-349, ICCE Digest, June (1992)

3)N.N. Nemeth, J.M. Manderscheid and J.P. Gyekenyesi, "Design of Ceramic Components with the NASA/CARES Computer Program," NASA TM 102369, April (1990)

4)A. Ghosh, C.Y. Cha, S. Vaidyanathan and W. Bozek, "Reliability Analysis of Cathode Ray Tubes," pp 508-510, Proc. of Society for Information Display, Volume XXIII (1992).

5)C.Y. Cha, S. Vaidyanathan, J. Trevelyan, "Elastic Stress Analysis Techniques for Assessing the Doming of a Cathode Ray Tube," pp 19-42 in Proc. 15th Int. Conf. on Boundary Element Methods in Engineering, Vol. 2, Computational Mechanics Publications, (1993).

6)C.Y. Cha, S. Vaidyanathan, S. Trevelyan, A. Adamski, W. Bozek, "A Comparison of Techniques for the Elastic Stress Analysis of a Cathode Ray Tube," Int. J. of Eng. Analysis with Boundary Element Method, 35-43 Vol. 13, [1] (1994).

7)C.Y. Cha, S. Vaidyanathan and J. Trevelyan, "Comparison of Convergence and Modeling Times for Cathode Ray Tube Stress Analysis with the Finite Element and Boundary Element Methods," Boundary Element Communication, 1-5, Vol. 6 [1] (1995).

8)C.Y. Cha, S. Iyer and J. Trevelyan, "Stress Failure Analysis of Neck Breakage of Cathode Ray Tube," pp 403-419, Proc. of 17th Int. Conf. on Boundary Element Methods in Engineering (1995).

OVERVIEW OF ASTM STANDARD ACTIVITIES* IN SUPPORT OF ADVANCED STRUCTURAL CERAMICS DEVELOPMENT

C. R. Brinkman
Metals and Ceramics Division
Oak Ridge National Laboratories
Oak Ridge, TN 37831-6154

G. D. Quinn
Ceramics Division
NIST
Gaithersburg, MD 20899

R. W. McClung
Consultant
901 Irwin Road
Powell, TN 37849

ABSTRACT

An overview is presented of the activities of ASTM Committee C-28 on Advanced Ceramics. This activity originated in 1986 when it became apparent that advanced ceramics were being considered for extensive use in applications such as advanced heat engines, heat exchangers, combustors, etc. in aerospace and energy conservation activities. These applications require optimum material behavior with physical and mechanical property reproducibility, component reliability, and well defined methods of data treatment and material analysis for both monolithic and composite ceramic materials. As new materials are introduced into the market place, these issues are best dealt with via standard methods. Therefore, a progress report is given describing activities of the five standard writing subcommittees who support the ASTM Committee C-28 effort. Accomplishments to date are given, as well as likely future activities, including a brief summary of joint cooperative efforts with international standard formulating organizations.

INTRODUCTION

The American Society for Testing Materials (ASTM) is the primary standards writing establishment in the United States for testing of materials, and is a private, nonprofit corporation for the development of full voluntary consensus standards on characteristics and performance of materials, products, systems, and services, and for the promotion of related knowledge. This is accomplished through the work of various ASTM committees consisting of volunteer experts, who following previously established regulations, generate a product (standard) that is widely recognized, accepted, and used. ASTM Committee C-28 on Advanced Ceramics was organized in 1986 when it became apparent that ceramics were being

* Research sponsored by the U.S. Department of Energy, Assistant Secretary for Energy Efficiency and Renewable Energy, Office of Transportation Technologies, as part of the Ceramic Technology Project of the Propulsion System Materials Program, under contract DE-AC05-OR21400 with Martin Marietta Energy Systems, Inc.

considered for many new high technology applications in the aerospace, biomedical, military, and automotive areas. These applications were viewed as being particularly demanding in terms of behavior requirements. Hence, it was important that standards be written for the production, inspection, testing, data analysis, and probabilistic design of advanced structural ceramics. An advanced ceramic can be defined as "a highly engineered, high performance, predominantly non-metallic, inorganic ceramic material having specific functional attributes," and includes both monolithic and composite ceramic materials. Accordingly, Committee C-28 was organized with various subcommittees as shown in Fig. 1 with task groups addressing specific technical topics under each subcommittee and writing appropriate standards. These standards may take the form of guides, practices, or test methods. Approximately one hundred and fifty members from industry, government laboratories, and academia serve on these various subcommittees. Additional participation and membership are always encouraged, especially as new technical issues are defined and additional expertise required. Subcommittee C-28.94 conducts liaison activities with the International Standards Organization's (ISO 206) technical committee on fine or advanced ceramics. A listing of completed or in draft form standards is given in Table I. Table I shows that fourteen standards have been completed since 1989 by the various subcommittees and that the level of effort is intensifying with time with a number of standards currently in the balloting and draft form while still others are being contemplated. After completion and approval of a standard, via the balloting process, it is published by ASTM in an annually revised book of standards[1]. The standard designation number gives the year the standard was approved. In addition to activities directly related to preparation of standards, the biannual meetings also sponsor technical workshops and symposia addressing state-of-the-art issues relevant to the work of the subcommittee. It is the objective of this paper to report progress in development and verification of standards by the various subcommittees for advanced ceramic materials. A brief summary of liaison activities associated with foreign and international standard formulating organizations will also be given.

PROPERTIES AND PERFORMANCE

Subcommittee C-28.01 on Properties and Performance has been developing standards associated with mechanical properties measurement for monolithic ceramics. It's standards are also applicable to particulate, platelets, whisker or discontinuous fiber-reinforced ceramic composites which are essentially isotropic and homogeneous.

As indicated in Table I, early work focused on issues associated with flexural strength determination both at ambient and elevated temperatures. This was done first, since historically the strength of ceramic materials has been determined by bend or flexural testing. While specimen costs are low in comparison to that associated with fabrication of tensile specimens and the test technique is relatively simple, many issues had to be resolved[2-5], including conduction of round robin tests, in order to develop these standards. Both of these standards have the restriction of being applicable only to materials tested under conditions of linear-elasticity. Since these standards were completed, results have been reported[6] of a large round robin test activity conducted using four hundred and fifty silicon nitride flexure specimens. Tests were conducted with specimens prepared in compliance with the standards and were conducted at room and elevated temperature by ten U.S. laboratories. One of the objectives of this work was to compare consistency of the coordinates of fracture initiation within the specimen relative to the loading fixture inner pins. Some indication of pin

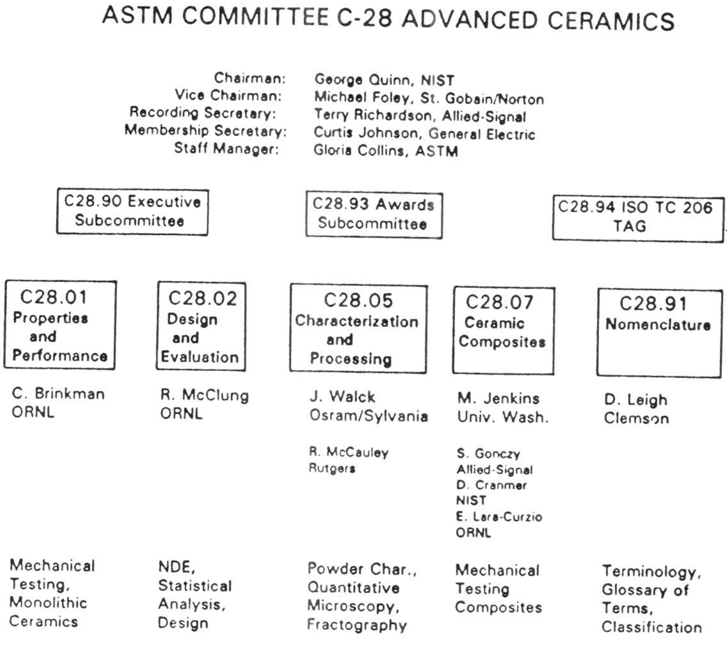

Fig. 1. Organization of ASTM Committee C-28.

Table I. Summary of Completed and Ongoing Standard Activities for ASTM Committee C-28 on Advanced Ceramics

COMPLETED STANDARDS

ASTM Annual Book of Standards, Vol. 15.01

Responsible Subcommittee	Designation	Title
91	C 1145-89	Definition of Terms Relating to Advanced Ceramics
01	C 1161-90	Test Method for Flexural Strength of Advanced Ceramics at Ambient Temperature
02	C 1175-91	Guide to Test Methods for Nondestructive Testing of Advanced Ceramics
01	C 1198-91	Test Method for Dynamic Young's Modulus, Shear Modulus, and Poisson's Ratio for Advanced Ceramics by Sonic Resonance
01	C 1211-92	Test Method for Flexural Strength of Advanced Ceramics at Elevated Temperature

02	C 1212-92	Practice for Fabricating Ceramic Reference Specimens Containing Seeded Voids
02	C 1239-94a	Practice for Reporting Uniaxial Strength Data and Estimating Weibull Distribution Parameters for Advanced Ceramics
05	C 1251-93	Guide for Determination of Specific Surface Area (of Advanced Ceramics) by Gas Adsorption
01	C 1259-94	Test Method for Dynamic Young's Modulus, Shear Modulus, and Poisson's Ratio for Advanced Ceramics by Impulse Excitation of Vibration
01	C 1273-94	Practice for Tensile Strength of Monolithic Advanced Ceramics at Ambient Temperatures
05	C 1274-94	Test Method for Advanced Ceramic Specific Surface Area by Physical Adsorption
07	C 1275-94	Practice for Monotonic Tensile Strength Testing of Continuous Fiber-Reinforced Advanced Ceramics with Solid Rectangular Cross-Section at Ambient Temperatures
05	C 1282-95	Test Method for Determining the Particle Size Distribution of Advanced Ceramics by Centrifugal Photosedimentation
91	C 1286-95	System for Classification of Advanced Ceramics

STANDARDS IN THE BALLOTING PROCESS

05	C xxxx	Test Method for Particle Size Distribution of Silicon Nitride or Silicon Carbide by X-Ray Monitoring of Gravity Sedimentation
01	C 1291	Test Method for Elevated Temperature Tensile Creep Strain, Creep Strain Rate, and Creep Time-to-Failure for Advanced Monolithic Ceramics
02	C xxxx	Practice for Measuring Ultrasonic Velocity in Advanced Ceramics
02	C xxxx	Practice for Fabricating Ceramic Reference Specimens Containing Seeded Inclusions
07	C 1292	Test Method for Shear Strength of Continuous Fiber-Reinforced Advanced Ceramics at Ambient Temperatures
07	C xxxx	Test Method for Flexural Properties of Continuous Fiber Reinforced Ceramic Composites
07	C xxxx	Test Method for Monotonic Compressive Strength Testing of Continuous Fiber-Reinforced Advanced Ceramics with Solid Rectangular Cross-Section Specimens at Ambient Temperatures
01	C xxxx	Test Method for Knoop Indentation Hardness of Advanced Ceramics
01	C xxxx	Test Method for Vickers Indentation Hardness of Advanced Ceramics
01	C xxxx	Test Method for Ultimate Strength of Advanced Ceramics with Diametrally Compressed C-Ring Specimens at Ambient Temperature

05	C xxxx	Practice for Fractography and Characterization of Fracture Origins in Advanced Ceramics
01	C xxxx	Test Method for Tensile Strength of Monolithic Advanced Ceramics at Elevated Temperatures
07	C xxxx	Test Method for Monotonic Tensile Strength Testing of Continuous Fiber-Reinforced Advanced Ceramics with Solid Rectangular Cross-Section at Elevated Temperatures
01	C/E	Test Method for Fracture Toughness of Advanced Ceramics

DRAFT STANDARDS

05	C	Test Method for Determination of Particle Size Distribution of Advanced Ceramic Powders by Electric Sensing Zone
05	C	Test Method for Determination of Particle Size Distribution of Advanced Ceramic Powders by Laser Light Scattering
05	C	Test Method for Determination of Total Carbon in Oxide and Nitride Advanced Ceramic Powders by Direct Combustion-Infrared Detection Method
05	C	Test Method for Determination of Total Oxygen in Carbide and Nitride Advanced Ceramic Powders by Direct Combustion-Infrared Detection Method
05	C	Test Method for X-ray Emission Spectrometric Analysis of Ceramic Powders

TASK GROUP PROJECTS-Draft Outlines

01	C	Test Method for Ultimate Strength of Advanced Ceramics with Diameterally Compressed O-Ring Specimens at Ambient Temperature
01	C	Dynamic Fatigue of Advanced Ceramics
02	C	Practice for Measuring Ultrasonic Attenuation Coefficients in Advanced Ceramics

CONTEMPLATED AND ON-GOING TASK GROUP PROJECTS

01		Test Method for Uniaxial Compression Strength of Advanced Monolithic Ceramics
01		Test Method for Uniaxial Tension-Tension Fatigue Strength of Advanced Monolithic Ceramics
02		Practice for Size Scaling of Uniaxial Strengths Using Weibull Statistics for Advanced Ceramics
01		Test Method for Diametral Compression Strength of Solid Cylinders for Advanced Monolithic Ceramics
02		Practice for Weibull Parameter Estimation Analyses that Enables the Pooling of Data from Multiple Specimen Geometries for Advanced Ceramics

02	Reference Standards Containing Laser-Drilled Holes in Advanced Ceramics
07	Method for Determining Ambient and Elevated Temperature Tensile Strength of Ceramic Fibers
07	Method for Determining Creep and Creep-Rupture Behavior of Ceramic Matrix Composites
07	Practice for Determining Tension-Tension Cyclic Fatigue Behavior of Ceramic Matrix Composites
07	Method for Determining Interfacial Properties in Ceramic Matrix Composites
05	Methods for Determining Density of Monolithic Ceramics
05	Test Method for Determination of the α/β Ratio of Silicon Nitride by X-Ray Diffraction

specimen interaction occurred in the high temperature fixtures compared to the room temperature results indicating that future changes in these standards may be desirable as more experience becomes available. Other countries and organizations have similarly been active in developing flexure testing standards including Japan, The Peoples Republic of China, Germany, France, and the Committee for European Normalization (CEN)[7].

Standards C 1198-91 and C 1259-94 enable the user to determine several dynamic elastic properties for advanced ceramics at ambient temperatures. In these methods resonant frequencies are measured while test specimens are vibrated at continuously variable frequencies. The resonant frequencies, dimensions and mass of the specimen are then used to calculate dynamic Young's modulus values. The dynamic shear modulus is found using torsional vibrations, and Poisson's ratio is then calculated from a knowledge of these two values. The former standard relies upon continuous excitation, whereas the latter uses the newer impulse excitation method.

Standards C 1273-94 and C 1291 deal with ambient temperature, ultimate tensile strength, and creep property determination, respectively, for monolithic ceramics subject to uniaxial loading. They are the result of advances made in recent years in solving problems associated with uniaxial specimen design, gripping, and alignment requirements[7]. The standard method for elevated temperature ultimate strength determination is expected to be approved in the near future.

Standards have been created for Knoop and Vickers hardness determinations. In conjunction with this, NIST is preparing two Standard Reference Materials (SRM's) which will have certified hardnesses. Figures 2 and 3 show Vickers and Knoop indentations using a 9.81N (1 kgf) and 19.61N (2 kgf) respectively. While either a 9.81 or 19.61N load is possible for the Knoop hardness measurement, the higher load will result in higher precision. Silicon nitride, Cerbec grade NBD-200 will be used for the Knoop SRM, while tungsten carbide with a cobalt binder, Kennametal grade KF 312, will be used as the Vickers SRM. Figure 4 shows a photograph of these SRMs which are expected to be available from NIST in the summer of 1995.

The need to determine the ultimate strength of tubular product forms using test specimens from the actual component when possible initiated work to standardize the diametrically compressed C-ring specimen test. The intent is to test specimens from the actual components with minimal surface preparation so as to determine strength behavior in the as-used condition. Whereas the C-ring test characterizes the strength of the external surface, the O-ring standard will characterize strength of the component internal surface. These two standards are aimed at ambient temperature characterization, however, their establishment will be followed by elevated temperature versions as a future effort.

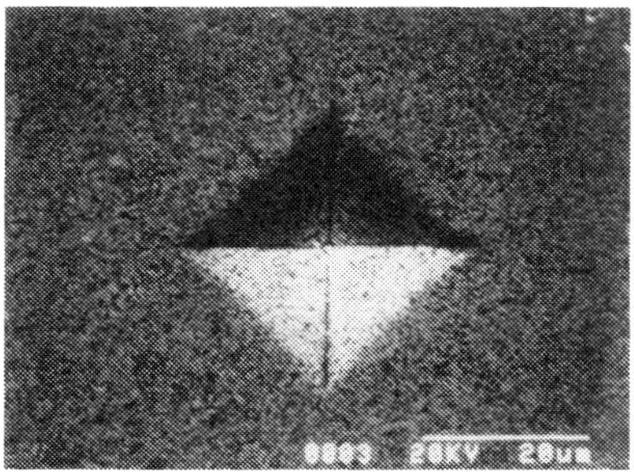

Fig. 2. SEM micrograph of a 9.8N Vickers indentation in KF-312 tungsten carbide.

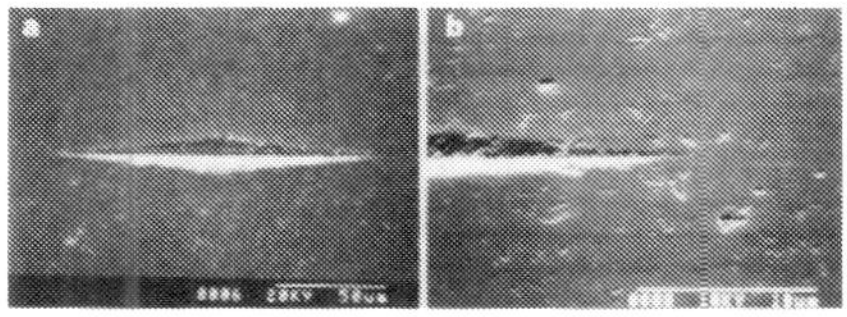

Fig. 3. SEM micrograph of a 19.6N Knoop indentation in NBD-200 silicon nitride.

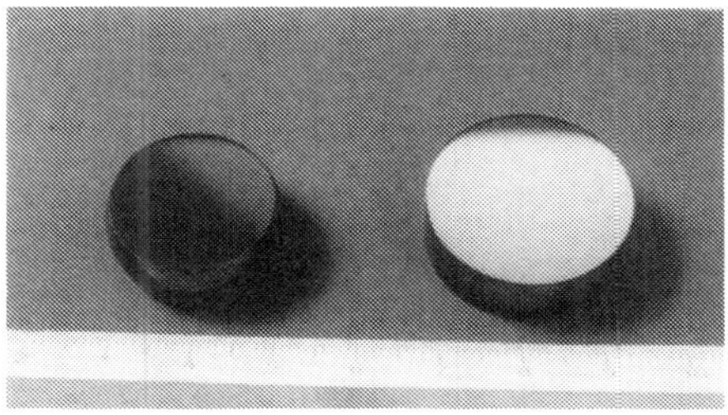

Fig. 4. Photograph of standard hardness reference materials.

DESIGN AND EVALUATION

Subcommittee C-28.02 on Design and Evaluation has two active sections. One on nondestructive examination and the other on probabilistic data analysis and design procedures.

The Section on Probabilistic Procedures (SPP), is concerned with preparation of standards related to such analytical procedures as bimodal analyses, fast fracture data analysis using Weibull statistics, size scaling of uniaxial strengths, pooling of data from multiple specimen geometries, and others. As a part of its activity, the SPP establishes cognizance and maintains awareness of the development of analytical and testing procedures in the United States (e.g., the NASA-developed CARES program[8,9] and the U.S.-DOE Ceramic Life Prediction Project[10-13] as well as in foreign countries (e.g. the European Weakest-Link Failure

Probability Prediction By Finite Element Postprocessors, "WELFEP", the International Energy Agency, IEA, tensile test project and other round robin test activities). A major effort was put into the development of C 1239-94a "Standard Practice for Reporting Strength Data and Estimating Weibull Distribution Parameters for Advanced Ceramics". This standard is restricted to uniaxial stress states, i.e., tensile, flexural, pressurized ring, etc., but is able to analyze multiple flaw populations and their effect upon the strength distribution. A maximum likelihood computer code (derived from CARES) is available on a floppy disk as an "adjunct" to this standard to perform the Weibull analysis.

In addition to the extensive work on the standard, the SPP also organized an international symposium on "Life Prediction Methodologies and Data for Ceramic Materials in Advanced Applications - A Basis for Standards" held in conjunction with the ASTM C-28 and American Ceramic Society meetings in Cocoa Beach, Florida in January 1993. The symposium addressed such topics as data and model development, life prediction methodologies, and prediction of the behavior of structural components. Papers have been published by ASTM in a special publication[14].

The Section on Nondestructive Evaluation (SNDE) has established a strong liaison with ASTM Committee E-7 on Nondestructive Testing to gain the benefit of the expertise available in that longstanding committee. An early activity of the SNDE was a technical review of E-7 standards (primarily in radiography, ultrasonics and liquid penetrants) to identify those that were considered (by the SNDE) to be relevant to advanced ceramics (although the original intent of E-7 was for application to metals). Some of the standards were recognized as directly applicable without change, some required minor modification for application to ceramics (without jeopardizing the original intent); others needed extensive change for applicability. Requests submitted to E-7 for the desired minor modifications have been well accepted and the changes have been incorporated and balloted by E-7. The more difficult extensive changes are in the current work program of the SNDE. A guideline standard that identifies the standards that are considered applicable to advanced ceramics (as originally accepted or modified) is C 1175-91, "Standard Guide to Test Methods and Standards for Nondestructive Testing of Advanced Ceramics." Another approved document that has been prepared by the SNDE is C 1212-92 "Standard Practice for Fabricating Ceramic Reference Specimens Containing Seeded Voids." Figure 5 is an example of detection of seeded voids in a flexural bar. Three new draft standards currently in the balloting process are "Standard Practice for Measuring Ultrasonic Velocity in Advanced Ceramics", "Standard

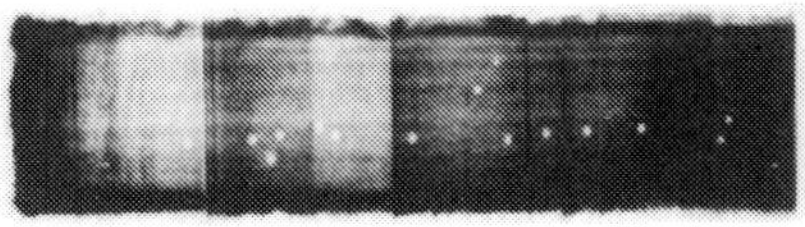

Fig. 5. Scanning Acoustic Microscopy image of voids in a silicon nitride flexure bar (courtesy of A. Vary, NASA Lewis Research Center).

Practice for Measuring Ultrasonic Attenuation Coefficients in Advanced Ceramics", and "Standard Practice for Fabricating Ceramic Reference Specimens Containing Seeded Inclusions." Actions are underway for potential standards on ceramic reference specimens containing laser-drilled holes, determining porosity in ceramics by ultrasonic velocity measurements and establishing radiographic equivalence factors (the last to be proposed upon completion for incorporation in an existing E-7 standard). An additional overview was recently published(15) of NDE techniques and standards development associated with monolithic and composite ceramic materials, and the reader is referred to this reference for additional information.

CHARACTERIZATION AND PROCESSING

Subcommittee C-28.05 deals with processing and characterization techniques. Areas of concern include techniques for physical, chemical, structural, as well as fractographic and microstructural analyses. There is considerable activity in powder characterization and several IEA round robins in support of powder characterization standardization have been very fruitful. Table 1 indicates that two standard practices have been completed in this area with an additional six standards currently either balloting or in preparation.

An example of the ongoing work by this subcommittee is the Standard Practice for Fractography and Characterization of Fracture Origins in Advanced Ceramics. This standard outlines an efficient and consistent methodology to locate and characterize fracture origins. While this standard is primarily applicable to analysis of fracture origins in mechanical test specimens, it is also applicable to component failure analysis as well. Figure 6 is an example montage of fractographs showing failure origins in silicon nitride uniaxial test specimens that had been subjected to tension-tension cyclic fatigue testing until failure occurred. A knowledge of fracture origins is considered to be a necessity in use of mechanical properties data in the evolving life prediction methodologies used in design. Recent publications give additional details concerning the origin of this standard(16) as well as results from an associated VAMAS round robin(17).

CERAMIC MATRIX COMPOSITES

Subcommittee C-28.07 on Ceramic Matrix composites was recently (1991) formed due to the rapid development of ceramic matrix composites that occurred during the 1980s and is concerned with standards issues related to mechanical property characterization of continuous fiber-reinforced ceramic composites (CFCCs).

Work on CFCCs is separated from that of monolithic or discontinuous reinforced materials due to their unique properties and behavior such as increased toughness, differences in failure mechanisms, and the fact that CFCCs typically consist of a brittle matrix with a stiffness similar to or greater than the reinforcing brittle fibers. These differences then define special test requirements associated with the necessity for appropriate specimen geometries and fabrication requirements, gripping test fixtures, specimen alignment, and specified optimal test conditions e.g. strain rates, temperature control, and measurement. Currently C-28.07 is comprised of a number of task groups addressing(18) standards associated with shear, flexural, compression, and elevated temperature tensile strength of CFCCs. As indicated in Table 1, these standard methods are in the balloting process, while work continues regarding draft standards for fiber strength, creep and creep-rupture, tension-tension fatigue, and interfacial properties in CFCCs. Forecasts project double-digit growth in the use of ceramic composites through the year 2000[19]. Standardization activities with resultant improved product reliability and reduced cost are expected to play a big role in developing the full market potential of this important class of materials.

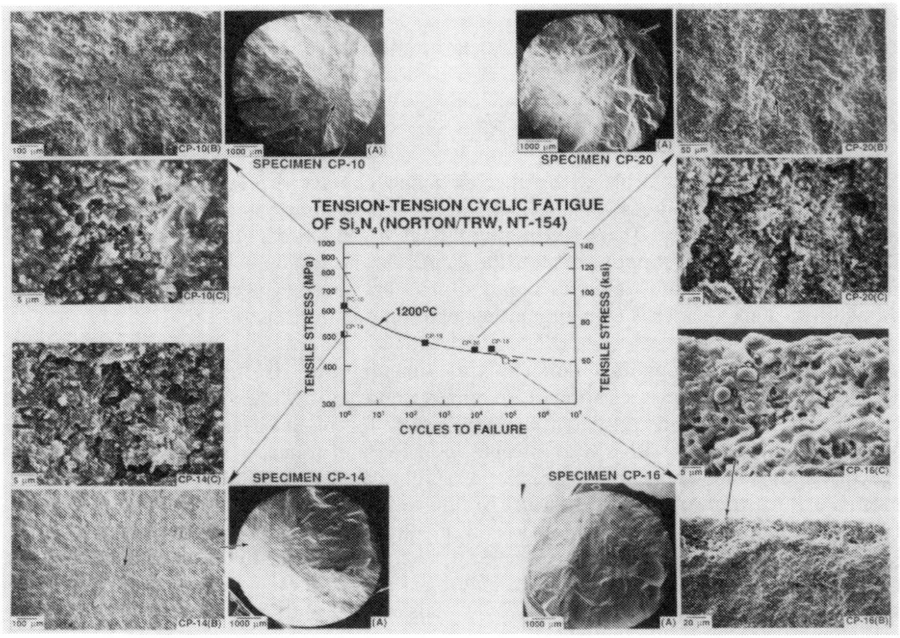

Fig. 6. Example of a fractography montage showing failure initiation site. Specimens shown have 3 fractographs each: overall fracture surface, fracture mirror, and origin close-up (courtesy of K. Liu, ORNL).

NOMENCLATURE

Subcommittee C-28.91 on Nomenclature provides the definitions of technical terms related to the development and characterization of advanced ceramics as used in the standards. As indicated in Table 1, Standard C 1145 provides this terminology. This standard continues to grow as new terms are added each year. A Standard System for Classification of Advanced Ceramics, C 1286-95, is another product of this subcommittee and provides a system by which ceramics, including composites, can be classified.

INTERNATIONAL STANDARDS ORGANIZATION (ISO) REPRESENTATION

Subcommittee C-28.94, ISO 206 Technical Advisory Group (TAG) is the U.S. representation to the ISO Technical Committee (TC) 206 on Fine (advanced) Ceramics. The aim of this committee is to provide international standardization in the field of fine ceramics powder, monolithic ceramics, fine ceramics based composite materials, and fine ceramic coatings. Currently ISO/TC 206 is comprised of members representing about 30 countries. C-28.94 provides delegates to attend meetings and to work on international standards and acquires U.S. consensus positions on all ISO TC 206 standards. Participation is important since ISO standards may affect international trade. Current collaborative effort includes work in four areas: Test methods for particle size distribution of ceramic powders, test methods for

flexural strength of monolithic ceramics at room temperature, test methods for hardness of monolithic ceramics at room temperature, and classification of fine ceramics.

CONCLUSIONS

Demand for advanced structural ceramics in the market place is expected to grow as these materials improve in consistency and reliability, and reduced cost. Standardized test methods are expected to accelerate use of these material as they become available and are used nationally and internationally. ASTM Committee C-28 on Advanced Ceramics has produced, as of this writing, fourteen high quality, technically rigorous consensus standards for processing, characterization, design, and evaluation of this class of materials. These activities have accelerated in recent years and many more standards are expected to be completed in the near future.

ACKNOWLEDGEMENTS

The authors gratefully acknowledge the assistance of Ms. J. Bozeman for preparation of the manuscript and Dr. D. R. Johnson for technical review. The authors wish to thank the members of Committee C-28 for their conscientious work in creating the standards listed in Table 1.

REFERENCES

1. 1995 Annual Book of Standards, Section 15, General Products, Chemical Specialties, and End Use Products, Vol. 1501. Refractories; Carbon and Graphite Products; Activated Carbon; Advanced Ceramics, ASTM, 1916 Race Street, Philadelphia, Pa. 19103-1187.
2. F. I. Baratta, "Requirements for Flexure Testing of Brittle Materials," U.S. Army Materials and Mechanics Research Center, Watertown, Mass., TR 82-20, April 1982.
3. F. I. Baratta, G. D. Quinn, and W. T. Matthews, "Errors Associated with Flexure Testing of Brittle Materials," U.S. Army Materials Technology Laboratory, Watertown, Mass., TR 87-35, July 1987.
4. G. D. Quinn, F. I. Baratta, and J. A. Conway, "Commentary on U.S. Army Standard Test Method for Flexural Strength of High Performance Ceramics at Ambient Temperature," U.S. Army MTL TR 85-21, Watertown, Mass., Aug. 1985.
5. G. D. Quinn, "Flexure Strength of Advanced Structural Ceramics, A Round Robin," J. Am. Ceram. Soc., 73, {8}, pp. 2374-84, 1990.

6. V. J. Tennery, K. Breder, and M. G. Jenkins, "A Study of Fracture Strength and Location Relations for 450 Silicon Nitride Flexure Specimen," to be published in the J. Am. Ceram. Soc.
7. C. R. Brinkman, G. D. Quinn, and R. W. McClung, "Development of ASTM Standards in Support of Advanced Ceramics Development-A Status Report," presented at the International Gas Turbine and Aeroengine Congress and Exposition, Cincinnati, Ohio, May 24-27, 1993, ASME Paper 93-GT-160.
8. N. N. Nemeth, J. M. Mandersheid, and J. P. Gyekenyesi, "Designing Ceramic Components with the CARES Computer Program," Ceramic Bulletin, 68, {12}, pp. 2064-72, 1989.
9. N. N. Nemeth, L. M. Powers, L. A. Janosik, and J. P. Gyekenyesi, "Time Dependent Reliability Analysis of Monolithic Ceramic Components Using the Cares/Life Integrated Design Program," Life Prediction Methodologies and Data for Ceramic Materials, ASTM STP 1201, C. R. Brinkman and S. F. Duffy, eds., Am. Soc. for Test. and Mat., Philadelphia, 1994, pp. 390-408.
10. P. K. Khandelwal, "Mechanical Properties and NDE of a HIPed Silicon Nitride," idem., pp. 127-140.
11. J. Cuccio, A. Peralta, J. Song, "Probabilistic Methods for Ceramic Component Design and Implications for Standard," idem., pp. 291-308.
12. A. D. Peralta, D. C. Wu, P. J. Brehm, J. C. Cuccio, and N. Menon, "Strength Predictions of Ceramic Components Under Complex Stress States," to be published in Proceedings of ASME Turbo Expo '95, Land, Sea, and Air, The 40th Gas Turbine and Aeroengine Congress, Users Symposium and Exposition, June 5-8, 1995, Houston, Tx.
13. P. K. Khandelwal, N. J. Provenzano, and W. E. Schneider, "The Development of Life Prediction Techniques for Structural Ceramics," idem.
14. C. R. Brinkman and S. F. Duffy, "Life Prediction Methodologies and Data for Ceramic Materials," STP 1201, Am. Soc. for Test. and Mat., 1994.
15. A. Vary, "NDE Standards of Ceramics and Ceramic Composites," ASTM Standardization News, Vol. 23, No. 1, Jan. 1995, pp. 40-45.
16. J. J. Swab and J. D. Quinn, "Results of a Round Robin Exercise on the Fractography of Advanced Structural Ceramics," Ceramic Engineering and Science Proceedings, Sept.-Oct. 1994, 18th Annual Conference on Composites and Advanced Ceramic Materials-B, The Am. Cer. Soc., 1994, pp. 867-76.
17. J. J. Swab and G. D. Quinn, "Fractography of Advanced Structural Ceramisc: Results from the VAMAS Round Robin Exercise," U. S. Army Research Laboratory, ARL TR-656, Dec. 1994.
18. M. G. Jenkins and E. Lara-Curzio, "Standardized Thermo-Mechanical Test Methods for CFCCs," AD-Vol. 43, Durability and Damage Tolerance, ASME 1994, pp. 191-211.
19. V. Smith, B. Deckman and D. Brueck, "Advanced Ceramics: Where Do We Go From Here?," Am. Ceram. Soc. Bull., Vol. 73, No. 12, Dec. 1994, pp. 49-52.

TRACING WARE WITHOUT USING BARCODES

Jeffrey Wagar
Decision Support Systems for Commercial Applications (DSSCA)
235 Johnston Road
Pittsburgh, PA 15241

Several overlapping papers on tracing ware, given by this author at the annual ACS meeting, have been combined into the paragraphs and tables shown below. These papers focused on: gaining the ability to trace problems back to individual casters, even without bar codes or casters marks; and the large cost savings achieved by this tracing ability.

Tracing Overreporting (to collect more Incentive Pay)

Figure 1

	Tom	John	Rick	Peter	Jim	Bob
Mon	$100	$100				
Tue		$150	$150	$150		
Wed	$120				$120	
Thu		$180	$180		$180	
Fri		$170				$170
Mon		$130				$130
Tue	$110	$110				
Wed		$140		$140		
	====	====	====	====	====	====
total	$330	$980	$330	$290	$300	$300

Casters are paid to meet a daily production quota based on the number of molds setup at their bench. Some casters may report 50 pieces produced, but place only 48 pieces on the ware trucks, relying on their ware being mixed into the vast stream headed to the dryers. A 2% casting shortfall in a plant with ten million in sales would cost $200,000 per year. Additionally, many thousands in incentive pay are collected for "phantom" ware. Lastly, Work-in-Process figures are inaccurate because of phantom counts. This leads to carrying extra WIP to offset shaky numbers, just when everyone is trying to get inventory costs under control.

Despite sending pieces into a continuous stream of ware, it is possible to trace overreporting. Incentive pay sheets are collected daily, then fed into a computer database. On Monday, if an item is short 4 pieces worth $25 apiece, those two casters, Tom & John, (out of six total casters in the department) who were producing that shape have $100 added to their [running] misreporting total in the database. Tuesday, another item (made by John, Rick, and Peter) is short 5 pieces worth $30 dollars apiece. Each of those 3 casters has $150 added to their misreporting total. If any of these 3 casters was involved in the $100 shortage from the previous day, then they would have a higher running total ($250) than the $150, $100, or $0 total of the other casters. Accumulate this data and produce the above table. The worker with the conspicuously high total, John, is present far too often (when misreporting occurs) to be a coincidence.

Tracing Shortages of Parts

Some factories assemble parts into a finished product. For example, Sanitaryware plants produce tanks & tank lids. Lids often have poorer yields than tanks themselves. Some casters either don't report the loss of lids (to get incentive pay for a complete tank & lid) or take lids from dryer trucks (and place them on their own tanks). If this practice cannot be traced to specific individuals, the factory will accumulate tanks (without lids) it cannot ship, ballooning WIP inventory and miss shipment deadlines. The method outlined above for tracing overreporting will also work for tracing chronic shortages of parts.

Tracing Flawed Ware without Caster Marks

A factory has sixty workers casting only four different models. Assume workers cannot put their mark on their ware because customers dislike seeing that mark. With many workers making the same model, how can the greenware inspectors determine who is submitting poorly finished or flawed ware? The trick is to divide the ware trucks into two groups by marking them differently — half might be painted blue, half marked with yellow.

Assume each of the six casters in Figure 2 are making the same model but at their own bench. Have the ware haulers split the casters into two different teams each day. For example, on Monday put all the ware from Tom, John, and Rick on the blue trucks; put the ware belonging to Peter, Jim, and Bob on the yellow trucks. Add up the total flawed ware from the blue trucks and assign Tom, John and Rick each one-third of that $300 total. Assign Peter, Jim, and Bob one-third each of the $135 total flawed ware from the yellow trucks. On Tuesday, have Tom, John, and Bob (instead of Rick) load their ware on the blue trucks, while Peter, Jim, and Rick load their ware on the yellow trucks. Again, assign each caster one-third of the value of the total flawed ware found on their team's trucks. Each weekday or week, vary the teams slightly. As time passes, the running totals for each caster will diverge. Those casters who consistently do a poor job will emerge with higher flaw totals than good casters.

Figure 2

	Tom	John	Rick	Peter	Jim	Bob
Mon	$100	$100	$100	$ 45	$ 45	$ 45
Tue	$150	$150	$ 50	$ 50	$ 50	$150
Wed	$ 80	$120	$ 80	$120	$ 80	$ 80
Thu	$110	$180	$180	$110	$180	$110
Fri	$ 90	$170	$ 90	$ 90	$170	$170
Mon	$ 75	$130	$ 75	$130	$ 75	$130
Tue	$110	$110	$ 85	$110	$ 85	$ 85
Wed	$100	$140	$140	$140	$100	$100
	====	====	====	====	====	====
total	$815	$1,100	$800	$795	$785	$870

This [extremely cheap] method will work even with large numbers of workers. If we indeed had sixty workers, we could simply divide the cast shop in half — all workers on the left hand side of the room use the blue trucks for a week, while casters on the other side of the room use the yellow trucks. Next week divide the shop front to back: workers in the front of the room use the blue trucks, et cetera. See which quarter of the room emerges with the highest flaw totals. Concentrate on that quarter (those fifteen casters) by splitting them into 3 teams of five, each using a different color truck. See which team turns in the highest totals. Eventually, you will pinpoint individual casters with casting problems or poor work habits.

Commercialization of Bioceramics

CHALLENGES IN CERAMIC DEVELOPMENT

Louis A. Serafin, Jr.

The fundamental purpose of this paper is to highlight some of the challenges encountered while bringing a medical grade structural ceramic to market. It focuses on the demands of the environment where the product will function while endeavoring to meet the Federal regulations that will govern its use.

-BACKGROUND-

The outstanding tribological properties of ceramics coupled with their inert behavior in adverse environments make them an ideal material of choice as an articular surface in the human body. Arthroplasty, for our purpose, will be defined as the replacement of human joints. Given the inevitable restrictions of space we will focus our ideas on the hip joint; a ball and socket system.

The normal hip in humans is composed of the proximal end of the femur bone which provides the ball of the ball and socket joint. It is coupled with the pelvis, composed of the ileum, ischium and pubis bones which form the acetabulum, or socket joint of the system. When the ball and socket are anatomically and physiologically married they are covered with a capsule in which the articular surfaces are bathed in synovial fluid.

Arthritis, which causes relentless pain, is seen as the deterioration of the joint space and integrity of the articular surfaces. It is one of the most common reasons senior patients seek medical attention. To alleviate the pain of arthritis physicians and scientists have developed mechanical joints. Physiological design, methods of attachment and installation are of utmost importance, however choice of materials leads us to the core of this paper.

-DEVELOPMENT-

The presently used systems have evolved from a single sided or hemiarthoplasty replacement of the ball made of alloyed metals to a complex total joint where all the articular surfaces are replaced with advanced materials. Our focus will be on the ball of the joint system. Time has proven that cobalt based alloys provide the patient with the best successful outcome. The American Society of Testing and Materials (ASTM) have registered multiple implant alloys, however most articular surfaces utilize F-799. The biocompatability of this product has withstood extensive testing in laboratory animals and decades of use in humans. The basic composition of this material is approximately 28% chromium, 6% molybdenum, 1% each of silicon, manganese and nickel with the balance being cobalt. The yield strength is > 820 MPa and the Rc is > 35. Forged material is machined into a ball which can be used as a surface replacement of the femoral head or be coupled to a conventional replacement inserted and fixed inside the femoral shaft. Our key analytical view will focus on the sphericity of the ball and microfinish of the surface. Since the balls are produced individually, the roundness and surface finish is often a reflection of machine integrity used in the manufacturing process. However, even under the most rigid, affordable manufacturing standards sphericity is only held between 100 and 500 millionths, and surface texture is dictated by carbides in the grains of the microstructure. Reducing the size and population of the carbides has improved surface quality but the best consistent surface finishes are measured in the 2 to 4 microinch range. Thus the option to use a ceramic surface seems an excellent choice.

Borides, oxides, carbides and nitrides were considered and narrowed to the oxides of aluminum and zirconium. Zirconia won as the ceramic of choice due to its strength, toughness and availability. It was proven via extensive testing that a ball could be economically produced which had a sphericity of less than 40 millionths of an inch and a surface roughness less than .5 microinches. This reflects an improvement of greater than five fold over the use of cobalt metals.

A proprietary, ultra clean zirconium powder was developed to meet self imposed standards. The Uranium Thorium (U-T) content was extensively studied at the University of Michigan, Phoenix Labs and Department of Public Health. It was determined that a total U-T content of 20 PPM or less was acceptable producing less than .8 REM exposure to the patient per year. Today routine powder analysis is less than one (1) PPM of total U-T content. The powder is isostatically pressed in rubber molds to form different size balls. The balls are exposed to a proprietary high-fire in a kiln which consolidates the balls to near theoretical density. This density reduces the flaw size and flaw population to near nil and allows an excellent quality control step before grinding. The balls are lapped into tolerance and the surface texture is brought to its final state with diamond finishing compound. The balls are again quality controlled and sent for a boring operation where a hole is placed in the ball. A taper angle is then ground in the hole which is highly controlled with air gauging.

-REGULATION

The Food and Drug Administration, Center for Devices and Radiological Health, controls the registration of medical devices in the United States. Guidances documents[1] to register ceramic femoral heads clearly challenge the manufacturers and evaluators of such products. The three most important categories of examination of the product before it can be brought to market are: (1) biocompatability, (2) structural integrity and (3) tribological properties.

(1.) Biocompatability of new implantable materials was evaluated under the Tripartite Agreement. The three main areas of examination are acute hypersensitivity, chronic hypersensitivity and mutagenicity. The calsigned powder fired particles were directly injected into the blood stream of laboratory animals. Implants were made from the as fired material and implanted in the muscles of the animals. The animals were examined both macroscopically and microscopically from an initial 24 hour period out to six months. In all cases while producing the experimental results in triplicate the animals never had any serious complications. No results were found that could be registered beyond the status of a slight irritant to the tissue. In many cases the results showed that the zirconium oxide was less of an irritant than the negative control implanted into the animals.

(2.) Structural integrity of the product required some of the most extensive testing. The balls were attached to conventional implants using a Morse taper system. In this situation balls could be gauged to different heights on the taper by controlling the depth of the bore into the ball. The tapers were mated, ceramic ball to metal trunnion and exposed to three test environments.

A. The balls were statically compressed on the trunnion to failure. No breakage was acceptable below 20 KN. A minimum of 5 balls of each size and neck length had to be tested and the average load to failure could not be less than 46 KN.

B. The same number of balls were exposed to fatigue testing to 10 million cycles. A static load of 1000 lbs. +/- 900 lbs. were cycled at 30 hertz on each ball. All specimens had to remain in tact during testing. At the completion of 10 million cycles the balls were taken to failure and no ball could fail at less than 20 KN.

C. The same number of balls were then surgically impacted onto the trunnions. They were then extracted from the trunnions and the pull-off force was measured.

(3) Tribological testing was related to surface profiling. Surfaces were measured for sphericity and found to be consistently < 30 millionths in roundness. A surface Ra range was established at .4 to .6 microinch finish. This data was acceptable for registration submission however extensive wear testing has been performed by multiple institutions. An excellent example can be found in the Journal of Biomedical Materials Research.[2]

-SUMMARY-

This zirconia ceramic ball has been used for more than four years on femoral components of the hip. Since the failure rate has been nil due to wear, it is impossible to predict the extended longevity that this ceramic ball may give over its metal counterpart. The wear tests and improvement in surface integrity should increase the life extensively. The retrievals that are available have shown no sign of compromise from the original manufacturing specification.

---

1 Guidance Document for the Preparation of PreMarket Notifications for Ceramic Ball Hip Systems; Orthopedic Devices Branch, Center for Devices and Readiological Health, U.S. Food and Drug Administration; CERBAL 95.1, Flash Fax (301) 443-9435.

2 Journal of Biomedical Materials Reasearch, Volume 25, Number 7, July, 1991 pp 813-828

Challenges in Bioceramics

Gary Fischman
Institute for Bioceramics
NSF-Industry/University Center for Biosurfaces
NYS College of Ceramics
Alfred University'
Alfred, NY

Bioceramics

When we talk about bioceramics, Larry Hench and June Wilson describe it as "the use of ceramics to improve the quality of life of humans"[1]. Although useful from the standpoint of ceramics for biomedical applications, if ceramics is to look at the market, the market is larger than suggested above. From this standpoint, I choose to define bioceramics more broadly: Bioceramics are materials that come into contact with a biological milieu of some sort. This definition allows us to consider a variety of situations including:

- Materials for Biomedical Devices - defined by the FDA as *any instrument, apparatus, or contrivance, including any and all components or parts, that were intended for use in the diagnosis, cure, treatment or prevention of disease in man or in other animals.*[2]

- Packaging for Medical Devices - although not in contact with a biological milieu, materials that give the sterilized device a shelf life is, in my opinion, an important biomedical material as it will have to undergo sterilization and then separate the biologics from the device being stored.

- Support Structures for Biologics - Support structures that can be used as a structure onto which organisms can be fixed for purposes such as filtration.

- Applications in Active Environments - Materials that will have to do their job in a bilogically adverse environment. This environment could be seawater, fresh water ponds, fish and water agricultural systems and so forth.

This seems to be a reasonable expansion as the properties of many of these materials are going to have to be similar (with possible exceptions in packaging).

From this standpoint the challenge to the ceramics industry will be to improve the quality and performance of the systems available as well as to bring about through new or better designed materials systems that are not yet available. Of course this will need to be done at a cost and reliability level commensurate with the market.

The reason that bioceramic markets have been described as above is because of the similarity of use and, therefore, likely cross markets among these groups. Imagine the environment that the material in a biological system as described above generally sees. Figure 1 schematically shows a material in a fluid environment which is what the materials would likely see in the worst case scenario of a biological environment.

Given the situation, there are many things that will happen just because there is an interface between the material and the environment, things such as

Dissolution of grain boundaries;
Dissolution of grains;
Diffusion of ionic species into the ceramic;
Formation of a passive or active coating;
Deposition of organic/inorganic species onto the surface;
Catalytic-heterogeneous reactions;
Ractions in the solution with ions from the material;
Incorporation of ions into biological molecules;
Change in solution chemistry - *e.g.*, changes in pH.

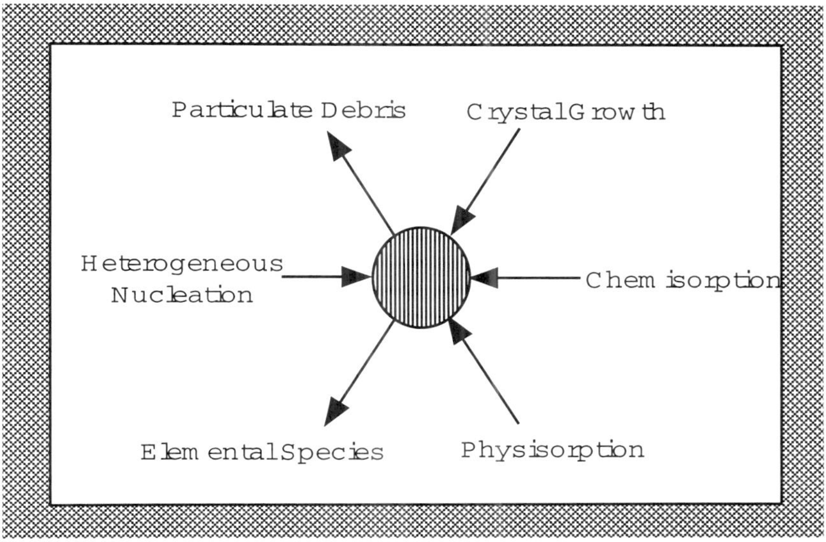

Figure 1: Schematic of the material environment system

Realistically, this means that if the area with the most stringent needs and controls is targeted, the other areas will quite likely also be targeted. Because of this, it is reasonable to look at materials for biomedical applications.

Biomedical Issues

Looking at materials for biomedical devices, we can look at basically four different issues: Material needs; FDA regulatory interaction; Liability and the "Materials Crisis"; and appropriate Marketing. Together these different issues will allow us to draw a picture of the directions we will need to go as an industry.

Material needs is really a statement of properties of pertinence. As described in an earlier paper[3], it seems to me to make distinctions among three levels of properties:

Crystal or phase properties;
Material Properties; and
Part or component properties.

In short, crystal properties are inherent to the crystal (or the phase if the material is amorphous), they tend to be directional and follow property descriptions such as those forwarded in Nye[4]. Material properties are the composite properties of a congregate mass of crystals, these are the properties we generally test for when using tests for engineering materials. Material properties depend upon the processing for the development of the general microstructure necessary to deliver the properties tested.

The part or component properties are simply extensions of the material property concept which we consistently use to test anyway. The part or component has a finite size and shape and has had to go through forming, firing and machining processes (and for biomedical materials often sterilization processes) that could change the properties of the material. Importantly, the properties of one section such as the surface wall or corner need not be the same as the properties of another such as the center of the bulk.

Since these three levels of properties are distinct, there is a need, at least at some point in the developmental process, to screen base materials (crystal properties), material properties such as 4 point bend as well as geometrically pertinent configurations.

Since the environment almost always contains a condensed fluid, a high concentration of mobile atoms, ions, proteins, organic entities from enzymes to organisms will be in close proximity to the material's surface. This suggests that the various effects described above will emerge and should be considered,

The FDA by way of its clearing devices for market and overseeing devices through post market surveillance and compliance is an integral part of the bioceramics challenge. By the way the FDA goes about clearing devices and not materials, it is clear that the FDA subscribed to the view of part properties as distinct from material and crystal properties. It could be said that the FDA

Clears <u>devices</u> for sale

does not clear materials for sale unless the material in and of itself is a device

requires tha devices (and therefore materials) be safe for implantation requires that devices work as intended

This means that basic biocompatibility issues must be resolved for each individual material and then the device must still be tested using the materials that will go into the marketed product.

One of the biggest challenges that faces the medical community and has caused material suppliers to shy away from the biomedical device market is liability. Materials are being withdrawn from the market due to high liabilities of materials in an industry that generally uses a low market share of these materials. This has caused a great deal of stir in the biomedical industry and creates an opening for industry that understands how to deal with the liability issue. "There must be greater emphasis on developing new and improved biomaterials so that the medical device industry can better meet the needs of clinicians and patients. In addition, government should take action to provide liability relief for raw materials suppliers."[5] As the need will not diminish, this will clearly be a fruitful marketplace.

Believe it or not, general marketing of ceramics materials and componenets in this field is practically nonexistent. This is a weak spot that can be strongly improved. By far the most prevalent materials in this field are organics. Surveying the Directory of Materials Suppliers furnished by Medical Device & Diagnostic Industry[6] gives the profile of material suppliers shown in Figure 2.

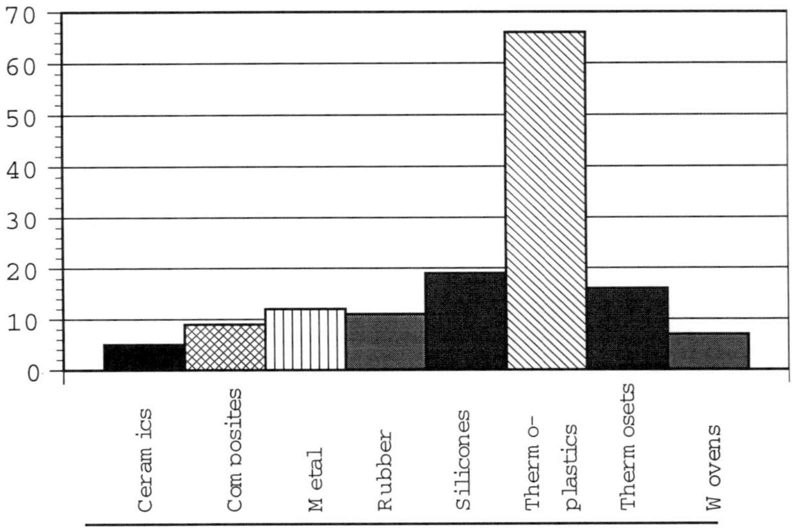

Figure 2: Compilation of Materials Supplier Data

Of the ceramic suppliers, one is a coater, one is an organic component manufacturer and two are machining companies.

Clearly, these results are skewed, there are companies that are developing ceramics for biomedical devices it is just that they are not clearly establishing themselves in the biomedical industry as a whole.

Deliverables

For there to be a challenge, there also must be deliverables to this challenge. There are exciting directions to move with respect to deliverables. The deliverables suggested here are my opinion of things that can be done. They are an educated guess based on my experience in teaching and learning. From the standpoint of new materials, the real challenge will likely be in the following areas:

Standard Biomaterials
Machinable-Healable materials
Rheologically invisible materials
Packaging

Standard biomaterials are the materials that have been discussed often in the past[7] which includes nearly inert, porous,bioactive and bioresorbable glasses and ceramics. There are needs in materials that have appropriate properties for the device in which they are intended. Beyond this, there are other directions that need to be developed.

One direction is that of machinable/healable materials. This suggests that a material can be machinable in the operating theater or in close proximity. This material would be a healable material, that is that before implantation or due to the environment, the material heals regaining the properties that it has lost through microdamage that has been created by machining.

Another direction is that of Rheologically invisible material and coatings. These are materials that have surface properties that make them invisible to the system. A good example of this would be a material that has the appropriate surface charge and smoothness that blood would mistake the material for epithelial wall tissue. Blood itself can be viewed as a colloidal suspension. The vascular walls are invisible to the blood until the vessel is cut, then the surface structure and chemistry changes and the blood in the region clots (flocculates).

Packaging is another strong possibility. New packaging materials are going to be necessary. Packages are not devices and therefore need not be held to the same controls as devices. Good packaging should:[8]

Guaranty sterility throughout the shelf life of the packaged device;

Be of GMP[9] quality (which parallels ISO 9000 guidelines);

Be of reasonable costs;

Be recyclable; and

Be environmentally sound.

Epilogue

In this paper I have suggested challenges in bioceramics. Although the term bioceramics was extended from its usual use to encompass similar systems, these challenges were focussed on the biomedical (and packaging) industry but can be extended into the industries for materials which must interface biological systems. Finally, I would like to put forth a few viewpoints on challenges in bioceramics.

There is a future in bioceramics. There are needs for materials with properties that are special for the particular applications involved. The need for materials will grow and the requirements which the materials will have to meet will become more specific. The development of materials will, in time, spur on changes in device design which, in turn, will activate further changes in materials. Ceramics should play a role in this situation. The devices themselves will continue to be made of components and the materials will likely be hybrid materials that will maximize the appropriate properties for that component.

One of the most daunting problems in the industry is the lack of protection in liability. Legislators are aware of this problem and the introduction of legislation such as the Biomaterials Access Assurance Act[5] has begun. This is a good sign which suggests that this problem may be short lived. Nonetheless, it will be in the best interest of the industry to put together an intelligent design and test network criteria that goes beyond what is currently in place. There is another challenge that does exist and this is where academia could get involved:

We must better understand the property relationships between crystals, microstructures and parts. Microstructural evolution and control and the relationship of the macrostructure to this and to the property set will have to be developed. A serious discussion of what the pertinent properties of these materials in these devices will have to ensue.

Having the information on properties and their development, we must design tests that minimize liability to the material producer. Although we cannot guard against everything, tests that evaluate the pertinent properties must be developed and carried out in a standardized form that will follow Good Laboratory Practices[10]. Facilities which develop test procedures based on better understanding of the material-system interactions and can test ceramics appropriately should be established.

Given the establishment of testing laboratories, protocols for not only testing but revisiting the validity of the testing for the environment-material system are essential. There needs to be an effort to show a constant improvement in understanding and development of these materials. The current models don't seem to be extremely predictive.

Putting all the pieces together, I am optimistic. I think that there is a place for ceramics as biomaterials and that biomedical applications will grow.

---

[1]LL Hench and J Wilson in An Introduction to Bioceramics, ed. LL Hench and J Wilson, World Scientific Press, Singapore (1993) page 1.

[2] E Horowitz and E Mueller, "Regulation of Medical Devices," IBID chapter 19.

[3] GS Fischman, "Characterization of Bioceramics," in Bioceramics: Materials and Applications, ed. G Fischman, A Clare and L Hench, American Ceramic Society, Westerville, OH 59-71 (1995).

[4] JF Nye, Physical Properties of Crystals, Oxford Press, England (196_).

[5] JS Benson and JW Boretos, "Biomaterials and the Future of Medical Devices," Medical Device & Diagnostic Industry 17 [4] 32-37 (1995).

[6]IBID 98-112.

[7]See for example: LL Hench and J Wilson, "Bioceramics" MRS Bulletin 16 [9] (1991) 62-74.

[8] from Medical Devices & Diagnostic Industry, August, 1994.

[9] Device Good Manufacturing Practices, FDA 87-4179, GPO 017-012-00330-3

[10] "Good Laboratory Practices (GLP) Regulations," Federal Register 52 FR 33768, September 4, 1987.

SOME ISSUES RELATED TO THE STANDARDIZATION OF IMPLANTABLE CERAMICS

Pietro Vincenzini
National Research Council
IRTEC,Faenza,Italy

ABSTRACT

A wide array of multidisciplinary problems are to be approached for a safe and reliable use of ceramics as implant materials. These include ethical, legislative, regulatory, scientific, technical and production aspects . The variety of available nearly inert, surface active and resorbable bioceramics, the several parameters that need accurate control, and , especially, the interactions between implant surface and living tissue are to be extensively evaluated and understood for the elaboration of standards. International activities underway on standardization of implantable ceramics are briefly reviewed and needs for future work outlined.

1-INTRODUCTION

When developing standards for ceramics to be implanted in humans, a number of issues -sometimes very specific issues- are to be taken into full consideration.

First of all, one has to consider that bioceramics -as well as other implantable materials- are a relatively new science. Indeed, although substitution of bone parts for repairing seriously damaged portions of a human body have been reported since the pre-Christian era -with either bronze or copper chiefly utilized- the first metal prostheses developed within a well organized programme directed towards achieving a practical and stable arthoplasty date to the late thirties. It was only at the beginning of the seventies that alumina ceramics was introduced, chiefly by Boutin (1972) following the extensive studies performed in the late sixties (1,2). In the same period, the first noticeable examples of bioactive glasses were developed (3). The above achievements may be considered as the first, truly systematic scientific approach from which the science of bioceramics springled out.

A second point that needs special consideration comes directly from the very special use of a material replacing a part of a living body. Ethical and juridical aspects indeed are involved in addition to the technical and scientific ones. A number of questions need to be given an answer in order to help in providing suitable guidelines for the operators involved (surgeons,engineers,scientists). The proposition of such guidelines is now being considered by recently established national and international committees (4-6).

Of course, the ethical principle by itself has not "the strenght of law". However, standards and laws must be inspired to the ethical commonly accepted principles (4-6). Furthermore, surgeons have their guidelines in the Ippocrate's code. Nevertheless, no written rules are at this time available on topical aspects such as:

- when may a biomaterial be utilized for test in a man?
- who decides if, when and how a prosthetic element may be implanted considering a specific material or a specific association of materials, choosing type, shape and composition from among different possibilities?
- will it be necessary to prepare a global normative or it is enough to formulate orientative indications about implant modalities of the prostheses?
- is it necessary to define standards on the way a surgical application of the different implants or it is wiser to leave to professional judgement to each individual surgeon?

A third point, somewhere related to the previous one, is the very scarce coordination existing at an international level among organizations devoted to the elaboration of standards. An harmonization of these activities should be seconded and strongly encouraged as well as the set up of a truly internationally accepted hierarchy of standardization organizations, as this will help to support decisions at government level for the elaboration of regulations and laws.

Presently, standard-related activities spread among a number of national and international bodies such as VAMAS (pre-standardization activities), ISO, CEN, ASTM, JIS, DIN, AFNOR, UNI etc. with frequently overlapping scopes and conflicting outputs. It would be appropriate that only one international reference body be recognized whose decisions shall be agreed with the widest international participation and consensus, and transfered to and adopted by each national standards institution.

A related problem is the different way standards are adopted inside national laws. For example, in Italy -as well as for other european countries- all the products classified as biomedical products, are not subjected to approval. This implies that a bad post operative result for a dubious use of materials, also if classified as biomedical products, falls on the surgeon. On the contrary, in Japan for every prosthesis or restorative device a specific approval by Ministry of Health and Welfare is needed, and in the USA from the American Food and Drug Administration. These shortcomings and undeterminations on appliable standards and laws have also relevant implications on the production industry and on the market.

Table 1. Main Requirements for an Implantable Material

- Biocompatibility
- Resistance to biological corrosion from hard and soft tissues, and biological fluids
- Suitable mechanical characteristics: Flexural, compression and shock resistance, yield point, elastic modulus, fatigue resistance, durability, tribological properties (fatigue,wear)
- Reliability
- Workability
- Acceptable cost

A last point to be considered comes from the large array of interdisciplinarity and multidisciplinarity of the activity concerned with implantable materials. Indeed, the several requisites for an implantable materials (Table 1) ask for a wide expertize from a number of different disciplines: operators in physical and biological sciences, designers and producers of materials and implants, and users (Table 2).

Consequently, any norms or standards shall arise from a suitable professional environment almost always involving diversified qualifications in several fields, and then make them known and accepted in practice by utilizers. This may pose sometimes severe problems when facing surgeons, who, on their side, are put under pressure by the several factors affecting the final result of a surgical operation such as pathologies present in the patient, choice of the most appropriate technique of implantation, inadeguate implant design, suitability of the chosen material as prosthetic constituent, etc.

Table 2. Disciplines Involved in the Formulation of Standards on Biomaterials

Categories	specialists
physical sciences operators	material scientists physics chemists
biological science operators	bioengineers biochemists
utilizers (surgeons)	dentists internists orthopaedists ENT operators neurophysiologists
medical field cooperators	veterinarians pathologists microbiologists cytologists
producers	Projectists Designers line production technicians

2-MAIN REQUIREMENTS FOR BIOCOMPATIBILITY OF IMPLANTABLE MATERIALS

Important requirements for a biomaterial to function in tissue reconstruction are biocompatibility, the capacity to support cell adhesion, the capacity to be remodeled and the capacity to exhibit physical forms with suitable mechanical properties.

Broadly speaking, two sets of characteristics are to be considered for an implantable biomaterial:

- Intrinsic properties of the material (chemical and physical), i.e. those pertaining to the exercize field of the material scientist and

- interaction of the material with the body tissues and fluids, this requiring imputs from a very broad range of disciplines.

All the above aspects may be ambraced by the general concepts of:

- Biotolerability
- Bioadaptability
- Biofunctionality

2.1 Biotolerability

Biotolerability is a global concept that involves a large number of single factors specifically correlated to the interactions with a living organism such as:

- Host response
- Bioadhesion
- Bioattachment
- System response
- Tissue induction (osteoinduction in bone)

It is evident that a living-tissue system involves an enormous amount of parameters, both biochemical and biophysical, many of which are unknown or at any rate cannot be controlled in a rational way, given the impossibility to always measure or verify them. The most convenient or obliged way of approaching the problem of interaction is therefore step by step, each step representing an increasing complex level of cell organization. Within each step, a number of fundamental aspects are taken into consideration which, however, can be defined only in a simplified experimental environment by: i) _in-vitro_ tests to assess leacheability in simulating physiological solution, and cytotoxicity and cytocompatibility with material in contact with cell cultures, and ii) _in-vivo_ tests to assess the reactive histological response (Host Response), mechanical adhesion to tissue (Bioattachment), the systematic aspects involved and the preservation of the chemical and chemico-physical characteristics of the material such as the biostability, the biodegradation, the bioresorption, the bioactivity, and the local and long-range chronic adverse effects.

2.2 Bioadaptability

Bioadaptability is a concept related to the degree of adaptation of a material substituted for a given tissue. The main factor involved is the mechanical one, i.e. the adhesive strength established by the material and the tissue ingrown or developed on its, shall we say, final form all around the material.

In the case of hard materials implantable into bone, the adhesive strength is often measured by the push-out test. This test takes into account in a comprehensive way a number of parameters all of which contribute to the mechanical stability of the implanted tissue, i.e.:

- The extent of contact (i.e. the ratio of the actual close contact area between tissue and the biomaterial and the overall surface area of the place in question)
- The yieldingness of the tissue formed close to the material
- The wettability of the biomaterial towards the surrounding tissue
- The liability of the implanted material to undergo cleavage by possible transformations occurring at the surface.

The materials to be exposed to cyclic friction, especially those utilized for joint coupling, must undergo tribological testing both at a mechanical level, by means of suitable devices (e.g. pin-on-disk or ring-on-disk), and at level of biological reactivity (biointeractivity) of powders towards host tissue and neighbouring tissue.

For all materials, but particularly for those placed into contact with blood, it is important to standardise the proteinic interaction tests, either for the protein lysis reactions or the protein absorption and accumulation onto the surface of a biomaterial. Another related aspect is the evaluation of blood compatibility, particularly of the kinetics of adhesion of the platelets and the modification of the trombogenetic factor.

With materials of some porosity it may become important to measure the extent of tissue ingrowth and outgrowth. It is also advisable that data should be supplied to detail the interactions of each material with drugs, radiations and systematic interactions.

The research activity on biocompatibility tests (a term which includes both biotolerability and bioadaptability aspects) clashes with requests of laws and rules. In this sector, ISO (International Standards Organization) established specific Technical Committees. The first committees involved for implants evaluation were those on dental implants (TC 106) and on surgical implants (TC 150). The common problems relative to biocompatibility tests brought to the formation of the Technical Committee " Biological Evaluation of Medical Devices" in 1989 (TC 194). This has been structured into 12 Working Groups each with an interest towards topics which are felt of major interest for biocompatibility (Table 3).
Technical reports including some proposals of international standards have yet been issued by this committee (7).

Table 3. Structure of ISO TC 184

BIOLOGICAL EVALUATION OF MEDICAL DEVICES	
WG1	Systematic Approach
WG2	Degradation Aspects
WG3	Animal Welfare
WG4	Clinical Investigation
WG5	Irritation and Sensitization
WG6	Effects on Blood
WG7	Cytotoxicity
WG8	Samples and Reference Materials
WG9	Mutageneticity and Carcenogeneticy
WG10	Systemic Toxicity
WG11	Implantation
WG12	Sterilization Residues

Whereas cytotoxicity (evaluating toxicity of leaching chemicals from test materials) was the concept concerning biocompatibility that was studied in ISO TC 194 WG-7, the activity of VAMAS/TWA 7 "Bioengineering Materials" approached biocompatibility by the viewpoint of cytocompatibility by assessing compatibility associated with electrochemical, micromorphological and physical characteristics of materials surfaces. A recent VAMAS/TWA 7 output (8) accounts for an international robin test carried out in the area of pre-standardization of biocompatibility test procedures by the cell culture method. Tentative procedures have been developed for using cell growth rate and cell adhesion rate tests to assess biocompatibility. The test results showed that the procedure yielded reproducible results for cell growth, but that, in the cell adhesion test, it was necessary to develop a procedure independent on the individual personal deviations in cell manipulation.

From a general viewpoint, it is to be noted that implantation into animals for biocompatibility tests (in-vivo tests) in addition to being time-consuming, expensive and not quantitative, is becoming uncreasingly undesiderable and socially unacceptable. Therefore the in-vitro test that gives quantitative evaluation with a reasonably short period of experiments, is expected to be promoted and improved for biocompatibility evaluation so that it could minimise the use of in-vivo test, although it could not perfectly reflect the in-vivo conditions.

2.3 Biofunctionality

Biofunctionality does not refer specifically to the material itself but mostly applies to the design criteria of a prosthetic device. "Rules" on the design criteria are required, since shape and surface finishing affect the answer of the host tissue. Such "rules" may also be adopted in preparing suitable computer aided design (CAD) programs by simulating tests of behaviour based on specific standards of design.

It is obvious that the required information may only be obtained through clinical trials,i.e. from results coming from implants in patients. This calls for a preliminary definition of minimum requirements or a quality mark for each prosthetic device and the standardization of the implant surgical technique for each kind of prosthesis. Obviously, this last requirement is especially difficult to face. Further requirements are availability of an informative data bank where all the important characteristics relative to the prosthesis, the clinical conditions of the patient and the adopted surgical technique should be collected. Useful information come from retrieval studies such as statistical information available on failures of hip implants (9), or the tumors experienced to come from systematic actions (10,11) due to continue release of ions or debris, and also the reconstructive difficulties due to incorrect design. The outputs coming from the EC Initiative "Concerted Action on Skeletal Implants", established in 1989 (12) and having among its primary goals to prepare methods of standardization and evaluation of retrieval skeletal implants, may provide an important source of data on the behaviour of various prostheses of different design, utilising a range of materials and associated clinical techniques, in view of the possible constitution of an internationally agreeable quality mark and hopefully of the related production procedures for each kind of prosthesis.

3-STATUS OF STANDARDIZATION OF IMPLANTABLE CERAMICS

3.1 Classification of implantable ceramics by type of fixation

Ceramics for surgical implantation may be classified according to the type of tissue response at the implant interface (Figure 1) (13) that realizes different means of achieving attachment to the musculo-skeletal system:

Nearly inert ceramics: a fibrous tissue of variable thickness not chemically or biologically bonded occurs between the implant and the tissue. In the case of a porous implant an interfacial bond forms because of ingrowth of tissue into pores on the surface or throughout the implant, provided the size of pores is greater than 100 to 150 µm.

Surface reactive ceramics: an interfacial bond forms between the tissue and the implant. Depending on material, strong tissue attachment to implant may be favoured.

Resorbable ceramics: designed to degrade gradually over a period of time and to be replaced by natural tissue. Main goal is to match the resorption rates to the repair rates.

Type of bioceramic	Type of fixation	Description of attachment	Materials
NEARLY INERT – Dense →	Morphological	Bone grows into surface irregularities by cementing the device, or by press fitting into a defect	Single crystal and polycrystalline alumina (PSZ and TZP zirconia under evaluation)
NEARLY INERT – Porous →	Biological	Bone ingrowth occurs, which mechanically attaches the bone to the material	Porous polycrystalline alumina, hydroxyapatite coatings
SURFACE REACTIVE	Bioactive	Attaches directly by chemical bonding with the bone	Bioactive glasses and glass-ceramics, hydroxyapatite
RESORBABLE	Resorption	Ceramics are slowly replaced by bone	Tricalcium phosphate, calcium phosphate salts, calcium sulfate

Figure 1. Classification of bioceramics according to type of fixation

Although all above materials, from nearly inert to bioactive, form an interfacial bond, the rate of bonding, the strength of the bond, the mechanism of bonding, and the thickness of the bonding zone differ for the various materials (Figure 2) (13). An example of the complexity of reaction stages of bonding bone to bioactive glass is shown in Figure 3 (14).

Such differences imply that specific requirements to be standardized

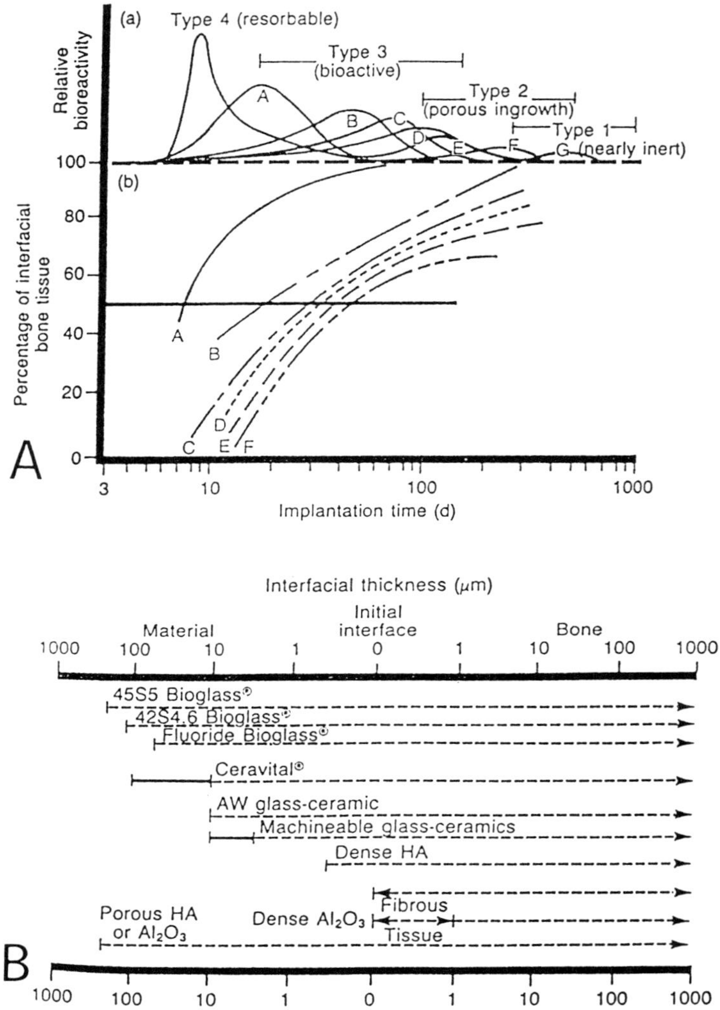

Figure 2. Bioactivity spectrum **A** and interfacial thickness **B** of reaction layer of bioactive implants or fibrous tissue of inactive bioceramics in bone (13)

depend on the various types of bioceramics and from their forecast use as bulk implants, coatings or composites, either in dense or porous form.

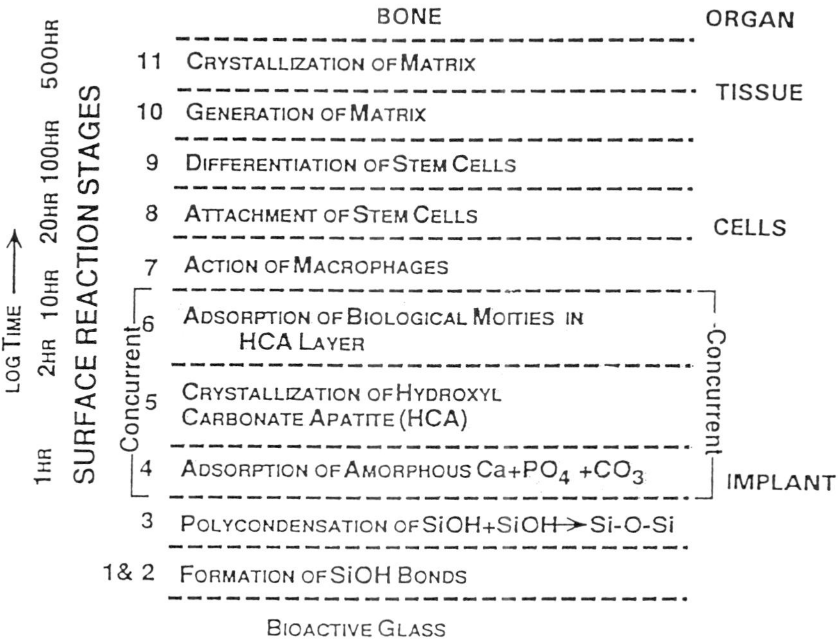

Figure 3. Reaction stages of bonding to bioactive glass (14)

3.2. Status of standardization of Advanced Technical Ceramics (ATCS)

It is noteworthy pointing out that standardization of testing methods and of requirements of implantable ceramics has not much progressed till now, and this is true also for the whole field of the so called Advanced Technical Ceramics (ATCs).

It is only in recent years that a remarkable activiy on standardization of test methods for ATCs has been started at an international scale. Main organizations involved are VAMAS (Versailles Project for Advanced Materials and Standards), ISO (International Standardization Organization), CEN (European Committee for Standardization), JISC (Japanese Industrial Standards Committee) and ASTM (USA). Other national organizations are engaged in the subject as well.

VAMAS activity in the pre-standardization of ATCs started in 1988 and it is specially directed to the standardization of test methods. Active in this field are VAMAS TWA 1 "Wear test Methods" and VAMAS TWA 3 "Ceramics", whereas VAMAS TWA 14 is involved in the "Classification of Advanced Ceramics" (Table 4). Work of the last committee is carried out in close contact with CEN TC/184, ASTM C28 and BSI RPM/13.

Table 4. VAMAS Activity in the Field of Pre-Standardization of Advanced Technical Ceramics

VAMAS Group	Specific projects
VAMAS TWA 1 "Wear Test Methods"	- Wear behaviour of selected ceramics - Standard format for organizing and reporting wear data - Interlaboratory wear volume measurement comparison - Compilation of standard wear test methods - Interlaboratory wear testing of inorganic coatings
VAMAS TWA 3 "Ceramics"	- Room temperature hardness - Fracture toughness at ambient and high T (IS/SEPB/IF) - Grain size - Surface cracks in flexure
VAMAS TWA 14 "Classification of Advanced Ceramics"	- Development of an international classification system for Advanced Ceramics, particularly nomenclature and terminology - Classification structure and mechanisms for system implantation

* Objectives of VAMAS pre-standardization activities are to undertake pre-standardization research on the reliability and reproducibility of test procedures for Advanced Technical Ceramics

CEN TC/184 "Advanced Technical Ceramics" was started in 1990. More than thirty new standards have been issued till now in the field of standardization of test methods of ATCs according to the tasks of the five Working Groups (Table 5).

ASTM C28, established in 1986, completed the first seven standards on testing methods for ATCs (C 1145-89 to C-1239 93) in the period 1989 to 1993. Others may have been issued during 1994 as many were at the level of balloting or draft outline (Table 6).

JISC activity on standardization of ATCs (called "Fine Ceramics" in Japan) started in 1981. In 1988 the Agency for Industrial Science and Technology (AIST) of Ministry of International Trade and Industry (MITI) set up a Special Committee for Standardization of New Materials within JISC. A basic guideline has been set up for 219 separate items for ATCs. Nineteen standards have been issued up to 1993 (R1601-1981 to R1618-1993). Drafts for additional six standards were issued during 1993, all concerning test methods on ATCs.

Table 5. CEN/TC 184 on Advanced Technical Ceramics Established in 1989

Working Groups		General Tasks
WG 1	"Classification"	- Development of a classification scheme - Terminology
WG 2	"Ceramic Powders"	- Development of test methods for assessing chemico-physical characteristics of powders and their behaviour under the processing cycle
WG3	"Monolithic Ceramics"	- Development of test methods for assessing physical and chemical physical properties of bulk ceramics at ambient and high temperature
WG 4	"Ceramic Composites"	- Development of test methods for assessing physical and chemico-physical properties of long fiber reinforced ceramic composites at ambient and high temperature
WG 5	"Ceramic Coatings"	- Development of test methods for assessing physical characteristics of ceramic coatings

Table 6. ASTM C-28 "Advanced Ceramics" Established in 1986

Sub-Committees		General Tasks
C28-01	"Properties and Performance"	- Mechanical testing ,monolithic ceramics
C28-02	"Design and Evaluation"	- NDE,statistical analysis,design
C28-05	"Characterization and Processing"	- Characterization,particle size
C28-07	"Ceramic Composites"	- Mechanical testing
C28-91	"Nomenclature"	- Glossary of terms
C28-94	ISO 206 TAG	- ISO/TC 206 coordination

ISO/TC 206 "Fine Ceramics" has been established quite recently. The proposed activity overlaps, of course, with the ones carried out by CEN, ASTM and JISC (Table 7). An effort should be commendable at an international level for establishing suitable links and an harmonized status among the different standardization initiatives. Discussions to overcome this problem are in progress.

Although all afore mentioned activities on standardization do not refer specifically to the biomedical use of an implantable ceramic, there is no doubt that several of the standard procedures developed

Table 7. Draft New Work Item Proposals for ISO/TC 206 (1994)

Draft New Work Item Proposals in the first stage
1. Classification of Fine Ceramics
2. Particle Size Distribution of Ceramic Powders
3. Flexural Strength of Monolithic Ceramics at RT
4. Hardness of Monolithic Ceramics at RT
Draft New Work Item Proposals in the second stage
5. Specific Area of Ceramic Powders
6. Tensile Strength of Monolithic Ceramics
7. Fracture Toughness of Monolithic Ceramics
Draft New Work Item Proposals in the next stage
8. Thermal Shock of Monolithic Ceramics
9. Elastic Moduli of Monolithic Ceramics
10. Grain Size of Monolithic Ceramics
11. Statistic Evaluation of Fracture Testing of Monolithic Ceramics
12. Absolute Density of Ceramic Powders
13. Density and Porosity of Monolithic Ceramics
14. Flexural Strength of Monolithic Ceramics at High Temperature
15. Thermal Expansion of Monolithic Ceramics
16. Terminology of Fine Ceramics

or yet under elaboration may also apply to standardization of implantable ceramics.

3.2 Standards specifically developed for implantable ceramics

Only few standards have been developed specifically for implantable ceramics. These concern primarily nearly inert ceramics (especially fully dense alumina implants and, in much lesser extent zirconia) and implants based on bioactive ceramics, mainly implants based on calcium phosphates (HA and ß-TCP).

3.3.1 Nearly inert ceramics

The use of high-density, high-purity (> 99.5%) alumina in various load-bearing implants such as hip joints and dental implants as well in other various clinical applications is widely known as from few decades. Therefore it is not surprising that alumina is possibly the only one example where standardization of test methods and of quality requirements for use in surgical implants have reached a satisfactorily, although not yet exhaustive, level of advancement.

Standards for alumina hip joints have been developed by DIN (DIN 58835), ISO (ISO 6474) and ASTM (ASTM F 603-83) in 1979, 1981 and 1983 respectively. A standard is also known from the former USSR. Available standard specifications for high purity alumina ceramics for

Table 8. Work Items and Standards Developed by JIS, ASTM C-28 and CEN/TC 184 Relevant to Implantable Ceramics

Title (abbreviated)	JIS	ASTM/C-28	CEN/TC184
Test Methods for Ceramic Powders			
-Impurities in Al_2O_3	CSJ	(C 573)	EN 725/1
-Impurities in ZrO_2	========	(C 705)	========
Test Methods for Monolithic Ceramics			
-Sampling and testing	========	========	ENV 1006
-Cracks by dye penetration	========	========	EN 623-1
-Density and porosity	(C2141)	(F77,C373)	EN 623-2
-Grain size	========	========	ENV 623-3
-Surface roughness	========	========	ENV 623-4
-Flexural strenght	R1601-81	C 1161-90	EN 843-1
-Elastic moduli	========	========	ENV 843-2
-Sub-critical crack growth	========	========	ENV 843-3
-Hardness	R1610-91	C	ENV 843-4
-Statistical evaluation of fracture testing	R16XX	C 1239-93	========
-Flexural creep	R1612-92	C	========
-Fracture toughness	R1607-90	C/E	========
-Tensile strength	R1606-90	C XXXX	========
-Compressive strength	R1608-90	(C 7773)	========
-Wear resistance	R1613-92	=========	========
-Tensile creep	R	C XXXX	========
Test Methods for Ceramic Composites			
-Tensile strength	========	C XXXX	ENV 658-1
-Compressive strength	========	========	ENV 658-2
-Flexural strength	========	========	ENV 658-3
-Shear strength (compression)	========	========	ENV 658-4
-Shear strength (3-point)	========	========	ENV 658-5
-Shear strength (double punch)	========	========	ENV 658-6
Test Methods for Ceramic Coatings			
-Thickness (step and cap grinding)	========	========	ENV 1071-1
-Thickness (grinding)	========	========	ENV 1071-2
-Adhesion (scratch)	========	========	ENV 1071-3

EN = European Norms   ENV = European pre-Norms

surgical implants are collected in Table 9. A revision of ISO 6474 issued in 1992, also reported in Table 9, accounts for the different conditions of use according to the effective mechanical load the implant has to bear during use (high- and low-load applications).
It also accounts for some details not previously considered such as the Fe_2O_3 content (possibly detrimental for corrosion resistance), and of the effect of excessive amounts of silica, CaO and alkaline oxides (which may affect strength). Also a reduction of the acceptable grain size has been felt important in order to minimize occurrence

of large flaws that may be critical to mechanical integrity.

An even more restrictive standard for alumina is needed for dental applications, as it has been stressed by the Working Group "Dental Implants" of ISO TC/106. ASTM Draft Specification F04.02.03.04 requires for the single crystal utilized in dental surgery a flexural strenght of 900 MPa.

Standardization of test methods and of key requirements not simply referring to the material but to be applied to a specific implant, such as load bearing capacity, friction behaviour in simulated physiological conditions, surface finish and sphericity of an alumina femoral head, is not yet accomplished. Some guidelines do exist, concerning, for instance, the dimensions, the machining quality and the mechanical quality of the ball-stem system which have been issued by FDA (15) and ISO. However, static and cycling (fatigue) tests, and pull-off tests proposed by FDA , recently have been the object of severe criticysm (16) deriving from the experimental observation that none of the available ball-stem combinations met the FDA requirements, although the same types of prostheses have been successfully implanted for many years.

While alumina has a history of about 20 years in clinical evaluation, zirconia is a relatively new material in the field of biomechanical applications. Consequently, there are different opinions and some doubts about its advantages and possible drawbacks.

Partially Stabilized Zirconia (PSZ) and Tetragonal Zirconia Polycrystals (TZP) are materials with very attractive mechanical properties, such as high strength and toughness. Nevertheless, not too much is known about the behaviour of zirconia for use in high loaded biological implants, such as, for example, hip joints. Information is still lacking about the long-term wear rate as well as the fracture behaviour of zirconia under fatigue cycling for both short- and long-term (over 20.10^6 cycles) in-vivo and in-vitro environments (17). Another related factor is the dubious long term structural stability of PSZ and TZP (tetragonal-monoclinic transformation) in a biological environment.

The Draft ISO/CD 13356 "Implants for surgery: ceramic materials based on yttria-stabilized tetragonal zirconia" issued in 1994 by ISO (ISO/TC 150 SC 1 N.200) and the Draft ASTM F04.02.03.14 "Zirconia ceramics for surgical implant applications", are the only two attempts presently available on standardization of zirconia bioceramics (Table 10). These specifications cover chemical requirements, allowable impurity levels, physical and short-term mechanical requirements. The amount of stabilizer and the acceptable level

Table 9. Standard Specifications for High-Purity Alumina Ceramics for Implants in Surgery. The values reported in the column New ISO are Coming from Revision of ISO 6474-1981

Properties	Units	ISO/DIN♥	USSR♠	New ISO♦	
				Type A	Type B
Density	g cm^{-3}	≥ 3.94	≥ 3.96	≥ 3.94	≥ 3.90
Bending strength	MPa	≥400	≥ 600	≥ 450	≥400
Compressive strength	Mpa	≥ 4500	≥ 4500	-	
Average grain size	μm	≤ 7.0	≤ 2.0	≤ 4.5	≤ 4.5
Impact strength	N cm^{-1}	≥ 40	-	-	
Microhardness	HV	≥ 2300	-	-	≥ 2300
Modulus of elasticity	GPa	≥ 380	-	-	≥ 380
Fracture toughness	Mpa $m^{1/2}$	-	5	-	-
Wear resistance (ring on disk)	mm^3 h^{-1}	≤ 0.01	≤ 0.01	≤ 0.001	≤ 0.01
Solubility corrosion in biomedia	mg m^{-2} day^{-1}	≤ 0.1	≤ 0.05	-	≤ 0.1
Chemical purity	wt%	≥ 99.5	≥ 99.5	≥ 99.5	≥ 99.5
SiO_2 + alkaline ions	wt%	≤ 0.1	≤ 0.5	≤ 0.1 ♣	≤ 0.1
Sintering additives	wt%	-	-	≤ 0.5	≤ 0.5

Note: Ceramic materials of type A are intended for implants at high load applications (*e.g.* bearing surfaces on joint replacement) and type B is intended for implants at low load application (*e.g.* maxillofacial and middle ear surgery).

♥ First generation specification according to DIN 58835, ASTM F603-83 and ISO 6474.

♠ Indications of N15.013-86 from GOST.

♦ Draft proposal under discussion as 2nd edition ISO/DP 6474 16.02.1987 to be edited in fail after 1992.

♣ Value obtained as sum of SiO_2+Na_2O+CaO+MgO+Fe_2O_3 with SiO_2 < 0.01, Na_2O < 0.01 and Fe_2O_3 < 0.01.

Table 10. Physical and Chemical Properties of Standard TSZP (ISO/TC 150SC 1 n.200)(Draft)

Property	unit	requirement
Bulk density	g cm^{-3}	≥ 6.00
Mean grain dimension	μm	≤ 0.6
Mean biaxial flexure strength	MPa	≥ 550
Mean 4-point-bending strength	MPa	≥ 900
Chemical purity (ZrO_2+HfO_2+Y_2O_3)	wt%	>99.0
Y_2O_3	wt%	4.8 to 5.4
HfO_2	wt%	≤5.00
Al_2O_3	wt%	<0.5
Other non-radioactive oxides	wt%	<0.5
Uranium oxides + Thorium oxides	ppm	<20

of radioactive impurities are also specified. Nevertheless, properties most strictly related to long-term use in load bearing implants are not considered.

3.2.2 Bioactive ceramics

Two classes of bioactive ceramics may be roughly defined:
i - Bioactive glasses and glass-ceramics
ii- Phosphate ceramics (HA and ß-TCP)
The two classes differ for the bonding mechanism to tissue.

Bioactive glasses and glass-ceramics

Several glass and glass-ceramics compositions have been developed which bond to the bone and in special cases also to soft tissue (13). The bond occurs by the formation at the implant surface of a biologically active hydroxycarbonate apatite (HCA) layer through a time-dependent, kinetic modification of the surface. The bond will not form if the rate of HCA formation is too low. The thickness of interface ranges from several μm to up hundreds of μm (Figure 2). The bond resists substantial mechanical forces.

Of the several factors which affect the interface strength that is one of the main parameters that needs to be standardized, we may consider:
i -Chemical composition. Intentional addition or presence of unwanted impurity may play a relevant role in formation and strenght of the bond. The positive or negative effect depend on type of addition/impurity and on actual glass composition. For example, although additions as little as 3wt% Al_2O_3 to the 45S5 bioglass, to low-alkali (0 to 5wt%) bioactive glass-ceramics Ceravital, and to A/W glass ceramic inhibit bone bonding, other kinds of glass-ceramics bond to bone even though alumina is present in the composition. However, in this case the Al^{+3} ions are incorporated within the crystal phase and do not alter the reaction kinetics of the material.

ii-Morphological factors. These are time-dependent factors and include:
- The change of interfacial area with time
- The progressive mineralization of the interfacial tissue.

These factors result in an increase of the elastic modulus of the interfacial bond and affect the shear strength per unit of bonded area. Also, it has been observed interfacial strength to be inversely dependent on the thickness of the bonding zone.
Few data exist to quantify these variables.

Chemical compositions of commercial glasses and glass-ceramics are normally covered by patents. Standards on test methods or on specification of chemica-physical and mechanical requirements for bioactive glasses and glass-ceramics are not yet available. The only one document on this matter is ASTM F04.02.03.06-Draft Specification for Glass and Glass-Ceramics Biomaterials. This draft specification defines the material requirements and characterization techniques for glass and glass-ceramic biomaterials intended for use as bulk, porous, or powdered surgical implants, or as coatings on surgical devices. The document defines and differentiates between glass, glass-ceramics, and bioactive glass and glass-ceramic biomaterials. A New Standard Specification for Glass and Glass-Ceramic Biomaterials in now under voting (March 1995) at ASTM (19).

Nearly dense HA and ß-TCP ceramics

The bonding mechanism of dense calcium phosphate implants, and specifically HA, differ substantially than for bioactive glasses and glass-ceramics. In nature HA constitutes the mineral part of skeletal tissues. Thus in the case of dense HA bone bonds to the implant through a thin layer of growing bone showing an almost perfect alignment with the apatite crystals in the implant (17,18). This type of bonding may be referred to as "direct bonding".

A consequence of this ultrathin (0.05 to 0.1 μm) bonding zone is a very high gradient in elastic modulus at the bonding interfaces between HA and bone. Of course, the different bond characteristics between HA and glass implants should also reflect on the performance of the implants, although the significance of this difference clinically is as yet unknown.

The main limitation of HA implants, also non-porous HA, is low mechanical strength and low reliability under tensile loads in physiological solutions. Consequently, in clinical practice bulk HA ceramics are specially indicated for use in small unloaded implants such as the middle ear, low-loaded porous implants where bone growth acts as a reinforcing phase, and dental implants with reinforcing metaL posts. On the contrary, the use of dense or porous HA as coatings for metal implants appears to be a very attractive area for applications.

Standards on HA ceramics primarily deal with factors linked to chemistry, crystallinity, impurity content and content in crystal phases, all factors that have been demonstrated to affect significantly the bonding characteristics and reliability of HA implants.
The following standards are available till now:
- ASTM F1185-88 "Standard Specifications for Compositions of Ceramic Hydroxyapatite for Surgical Implants". This standard includes chemical

composition, the content of mineral phases and the admissible impurities in HA powders. One important standard restriction is the limit of at least 95% for the content in HA, since this appears presently the only way to ensure long-term chemical resistance of the implant, although recent experimental evidence suggests an even more restrictive limit of 98% for the minimum content in HA phase.

- AFNOR S94-066 (August 1993). Provides the precise determination of the Ca/P ratio of calcium phosphate powders and deposits. Precise conditions are also given for an HA to be considered stoichiometric, i.e. absence of ß-TCP, $Ca(OH)_2$ and calcium oxide.
- AFNOR S94-067 (August 1993). Standardize the qualitative and quantitative determination of foreign phases present in the powders, deposits and ceramics based on HA.
- AFNOR S94-068 (August 1993). Standardize the determination of crystallinity and of the apparent size of the apatite crystallites in powders, deposits and ceramics based on HA.

For ß-TCP the Standard ASTM F 1088-87 involves chemical cmposition (stoichiometry of the ß-TCP phase, maximum allowable limit of heavy metals) and crystal structure confirming also for ß-TCP a minimum phase purity of 95%.

Several draft specifications have been issued by ASTM in recent years concerning calcium phosphate based bioceramics. Draft Specification F04.02.03.08 defines terminology to be used in analyzing the crystallinity of calcium phosphate ceramics, whereas various draft specifications refer to use of calcium phosphates as coatings to surgical implants. These include:

- Draft Specification F04.02.03.03. Defines a number of terms related to calcium phosphate coatings, specifies chemical and/or crystallographic requirements of the coating, outlines the various physical and mechanical test methods to be used to characterize the coating, and , where possible, relates these techniques to existing ASTM standards.
- Draft Specification F04.02.03.05. Refers to calcium phosphate powders to be used in the preparation of biocompatible coatings.
- Draft Specification F04.02.03.10. Defines physical requirements such as thickness, surface roughness and uniformity, porosity, density, colour and morphology.
- Draft Specification F04.02.03.11. Is on test methods for determining the mechanical requirements (tension, shear, and bending and shear fatigue test).
- Draft Specification F04.02.03.12. Concerns the environmental stability of coatings by in-vitro evaluation of the dissolution rate.

The following documents concerning calcium phosphate coatings are being voted at ASTM (March 1994)(19):

- New Standard Specification for Evaluation of the Environmental Stability of Calcium Phosphate Coatings
- Revision of F1501-94, Test Method for Tension Testing of Calcium Phosphate Coatings.

4-SUMMARY AND CONCLUSIONS

Standardization of implantable ceramics appears quite unsatisfactory nowdays. However, we can be moderately optimistic that at least some of the major areas related to standardization of biocompatibility testing will be satisfactorily approached within ISO TC 184.

More critical appear the perspectives about a quick solution of the standardization of the several aspects linked to bioadaptability and biofunctionality performance of prostheses. This is a focal point also because of the strict links existing among standards and production and marketing aspects.

One of the most relevant items is assuring long duration. Present prosthetic devices suffer for lifetime problems. Depending on the specific application, troubles may come, for example, from a basic mismatch between the implant and the adjacent bone at region of fixation (micromovement effects) common in loaded implants made of bioinert ceramics, or from interfacial instability in bioactive implants because of degradation processes.

Available statistics of prostheses lifetime show that revision surgery will undergo an impressive excalation during the coming years with a resulting heavy effect on our health-care system. Studies now available on the tissue-implant interface of retrieved implants provide an important source of data on the factors contributing to implant failure, thus allowing an upgrading of novel prototype prostheses, assisting in harmonizing protocols for implant retrieval analysis, and contributing in establishing new standards on materials and devices. All these aspects are relevant to the objective of assuring long-term performance, a property that presently cannot be guaranteed especially for bioactive ceramics and coatings.

Standardization of test methods for a number of important requisites such as interfacial bond strength, toughness, scratch-hardness and fatigue life under realistic physiologic loads, among several others, remain among the priorities for standardization.

Obviously, most of the problems related to prostheses durability, hence to standardization, may be perspectively solved by improving current technology and exploring new directions towards a second-generation prosthetic devices.

Indeed, the full potential of bioceramics appears far from being entirely exploited. It may be reasonably expected that in the years to come an approach to implantable ceramics rooted on most intensive and appropriate use of biotechnology will be pursued. This will imply coupling of materials science with sensor technology, information processing and genetic engineering, also at the purpose of providing a certain degree of "intelligence" to materials and devices. Biomolecular manipulation of composition, refining the engineering of microstructure and surface chemistry, along with the incorporation of living cells, cellular components and cellular products, and synthetic components modeled on cellular biosynthesis will hopefully open the horizon to new products and applications hard to be imagined at present.

These foreseable advancements stress even more for the need that conditions be established for a coordinated hierarchical asset of standardization activities, which will enable us to cope with present and incoming demand for a reliable network of guidelines and standards for bioceramic implants. This need also couples with our moral responsibility as materials scientists that no new materials and devices be proposed in expanded clinical use before long-term reliability and full safety for the patient are assured at an acceptable level.

REFERENCES

1- S.F.HULBERT, Biomaterials-The Case of Ceramics, in "Use of Ceramics in Surgical Implants", S.F.Hulbert and F.A.Young Eds., Gordon and Breach Science Publishers, New York,1969

2- J.J.KLAWITTER,S.F.HULBERT,C.D.TALBERT,C.T.FITTS, Artificial Bones, in "Use of Ceramics in Clinical Implants", S.F.Hulbert and F.A.Young Eds., Gordon and Breach Science Publishers, New York,1969

3- L.L.HENCH,R.J.SPLINTER,W.C.ALLEN,T.K.GREENLEE,Jr.,Bonding Mechanism at the Interface of Ceramic Prosthetic Materials, J.Biomed. Mater. Res. **2**(1972)117-41

4- S.SALIA, Ethical Issues in "Biomaterials Research and Practice", Proc. of Fourth Biomaterials Congress, Berlin 1992 PD 1.1

5- S.B.RAO, Ethical Considerations in Biological Research:Perspectives from Developing Country, Proc. of Fourth World Biomaterials Congress Berlin,1992 PD 1.2

6- S.SALIA,P.SALIA, Ethical Issues on the Use of Animals in the Testing of Medical Implants, Proc. of Fourth Biomaterials Congress, Berlin, 1992 PD 1.3

7- M.PERREN, International Standardization in the Field of Biocompatibility Testing, Proc. 10th European Conf. of Biomaterials, Davos, Switzerland,1993,p.108

8- VAMAS Bulletin N.17, January 1994

9- I.GUALTIERI,S.GAGLIARDI,M.HENDRIKS, Results on the first 100 hip joints in bioceramics, Biomateriali **1**(1988)p.33-38

10-R.G.KAMMULA, Regulatory Requirements for the Carcinogeneticy Evaluation of Long-Term Implants, Proc. of the 19th Annual Meeting of the Society for Biomaterials,Birmingham,Usa,1993,p.226

11-M.NAGASE,M.CHIGIRA,E.UDAGAWA, Toxicity of Silico-Containing Calcium Phosphate Glasses Demonstrated in Mice, Biomaterials **13**(1992)p.172-175

12-EC Initiative "Concerted Action on Skeletal Implants, First Meeting 13th December 1989,Bordeaux,France

14-L.L.HENCH, The Theory of Bioactive Bonding, in "Materials in Clinical Applications", Proc. 8th CIMTEC-World Ceramics Congress and Forum on New Materials, P.Vincenzini Ed.,Techna srl,Faenza,Italy,1995 p.331-342

15-Guidance Document for the Preparation of Premarket Notifications for Ceramic Ball Hip Systems, FDA,April 27,1989

16-U.SOLTESZ,R.SCHAFER,D.SIEGELE,B.THIELICKE, Strength Testing of Ceramic Hip Joint Heads - Criticism of the ISO and FDA Procedures, in Proceedings 8th CIMTEC-World Ceramics Congress and Forum on New Materials, P.Vincenzini Ed.,Techna srl,Faenza,Italy,1995,p.583-590

17-M.JARCHO, Calcium Phosphate Ceramics as Hard Tissue Prosthetics, Clin. Orthop. Relat. Res. **157**(1981)p.259

18-M.OGISO,H.KANEDA,J.ARASAKI,I.TABATA,T.HIDAKA, Epithelial Attachment of Bone Tissue Formation on the Surface of Hydroxylapatite Ceramics Proc.1st World Biomaterials Congress, Band,Austria,1980,p.415

19-W.R.LACEFIELD,The University of Alabama at Birmingham, USA (Private Communication)

PROOF-TESTING OF CERAMIC HIP JOINT HEADS

H.G. Richter, M. Wimmer, U. Kemmer, G. Willmann
CERASIV GmbH, Plochingen, Germany

Abstract

Alumina hip joint heads have been successfully used for hip joint heads in total hip joint replacement for over two decades. Among the causes for revision surgery ball head breakage is almost neglegible. Nevertheless, in order to add an additional mechanical safety margin to ceramic hip joint heads a proof-testing system for ceramic hip joint heads has been developed where the load to the hip joint head is applied by liquid pressure. The determination of the proof-stress level is based on the in-vivo failure rate.

Introduction

Total hip joint replacement has been a successful orthopaedic surgery procedure for over two decades. Normally, in a total hip joint system several materials are utilized: the stem normally is manufactured of metals (Ti-alloy, CoCr-alloy), the cup in most cases is made of Polyethylene (PE), to a smaller extent there are also cups being used made of metals or of alumina. For the artificial hip joint head alumina has proven to be extremely useful due to its excellent wear behavior in combination with polyethylene cups, and with even better results with alumina cups.

Despite of the high reliability level there can still occur problems which make revision surgery on artificial hip joints necessary. Several causes for revision surgery are listed in Table 1. Aseptic loosening is by far the most severe revision reason. Aseptic loosening can be loosening of the cup caused by the body reaction to wear debris originating in the PE or metal cup. Loosening of the stem mostly is caused by bone cement degradation. Bone cement is used for fixation of the stem in the femur.

By far the least reason for revision surgery is fracture of a ceramic ball head. The number given in Table 1 is based on over 20 years of clinical experience and means that in the average only one ball head out of 10000 implanted ceramic ball heads breaks in-vivo.

Table 1: Reasons for revision surgery in total hip joint replacment

Aseptic loosening	up to	20 %
Bone cement degradation	up to	10 %
Infections	up to	1 %
Ceramic ball head fracture		0,01 %

In order to improve this already excellent rating several measures were undertaken in recent years: the alumina density has been increased from 3.9 to 3.97 g/cm^3, the mean grain size has been reduced from 7 µm to below 3 µm, the strength, measured on a specially shaped 3-point-bending specimens could be increased from around 400 to above 700 MPa. This advancements would not have been possible without introducing clean room technology.
It was possible to increase the performance level of alumina ball heads but it is not possible to evade the material inherent strength statistics by these measures. There is still the question how the ball heads at the low and of the strength distribution can be eliminated.

Several non-destructive test methods were investigated for this purpose including X-ray inspection, ultrasonic analysis and neutron diffraction. In the end proof-testing was chosen because it turned out to be the most efficient and reliable method.

In the best case X-ray and ultrasonic methods would find parts with flaws without assessing the relevance of the detected flaw to the strength of the component. In contrast to these methods the proof test makes an assessment of the strength relevance of the detected flaw at the moment of finding.

In introducing any method of non-destructive testing one is making the assumption that imperfections in the balls are the only reason for in-vivo

In introducing any method of non-destructive testing one is making the assumption that imperfections in the balls are the only reason for in-vivo failures. It is known from experience that this is not the case. Accidents which lead to overloading have to be made responsible to an appreciable extent for in vivo-failures. Nevertheless, eliminating the balls that are at the lower and of the strength distribution would provide an additional safety margin to the ball.

Proof-Testing Method

The theoretical basis for proof-testing was given by Evans and Wiederhorn [1]. The basic principle of proof-testing is based on the fact that a component breaks at a stress σ_p when the stress intensity factor generated at the worst flaw, K_{ip}, is higher than the critical value K_{Ic}

(1) $$K_{Ip} = \sigma_p \cdot \sqrt{a_o} \cdot f \geq K_{Ic}$$

where a_o is the length of the worst flaw.

When the surviving component ist used at a lower stress σ_s the initial stress intensity factor is

(2) $$K_{Is} = \sigma_s \cdot \sqrt{a_o} \cdot f \leq K_{Ic} \cdot \frac{\sigma_s}{\sigma_p}$$

Applying a stress to a hip joint head which resembles the stress distribution in service but exceeds it in level at every point weak hip joint heads can be eliminated at a predetermined level.

Application of proof-testing to ceramic hip joint heads

There were several attempts undertaken in the late seventies [2,3] to apply the idea of proof-testing to hip joint heads. But there seems to be no publication on proof-testing of hip joint heads on a production level.

The crucial step in setting up a valid proof-test for hip joint heads is how to generate the proof-stress. The proof-stress has to copy the loading situation which is given by pressing the ball onto the conical end of the stem. In the proof-test test it is not possible to use a metal stem because it is not possible to remove the stem from the ball without damaging the ball. For the present purpose a device was designed by which the ball head can be stressed by liquid pressure (see Fig. 1). Using this principle it is possible to meet the requirement that exactly the same area of the conical bore of the ceramic ball has to be pressurized by the liquid that can be in contact with the conical end of the stem. Pressures up to 1500 bar are necessary for testing alumina hip joint heads.

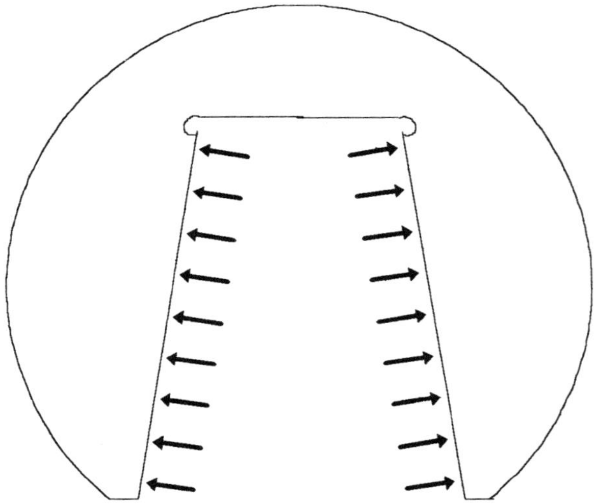

Fig. 1: Schematic illustration of the loading principle

Efficiency of proof-testing ceramic hip joint heads

The efficiency of the proof-test set-up was tested as follows:

1. The strength distribution according to ISO 7206-6 was measured on 120 balls
2. The distribution of pressures necessary to fracture hip joint heads in the proof-test setup was measured on 120 balls.
3. A third set of balls was proof-tested using at a pressure level corresponding to a failure rate of 5 %.
4. The hip joint heads that survived this proof-test were tested for resistance to static load according to ISO 7206-5

The results are illustrated in Fig. 2: The line represents the strength distribution of the not proof-tested set, the points give the strengths of the proof-tested balls. It is obvious that the fracture load distribution of the proof-tested set is sharply truncated as it is predicted by the proof-test mathematics.

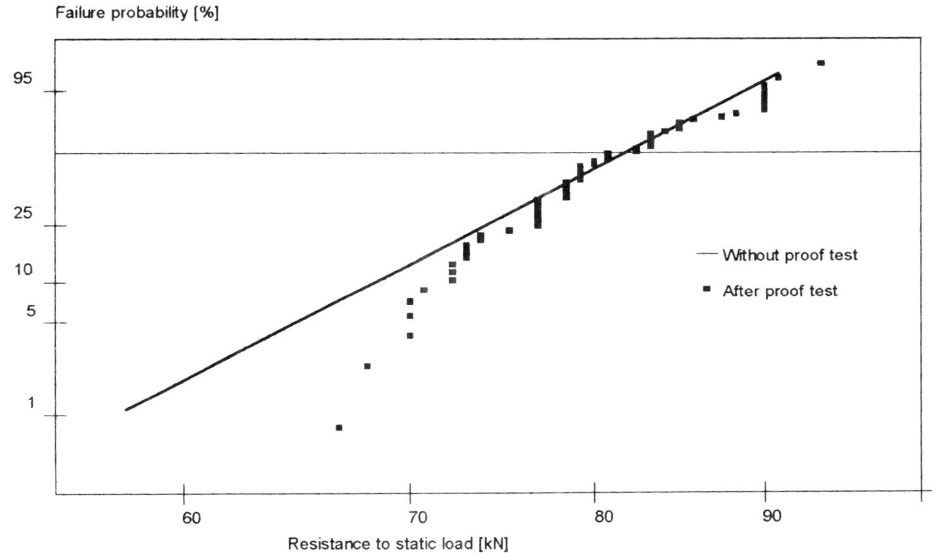

Fig. 2: Resistance to static load (ISO 7206-5) without and after proof- testing

It can be concluded that by the presented method of applying the proof stress weak hip joint heads can be eliminiated at a preset failure probability or load level.

Determination of the proof-stress level

The other important step in setting up a proof-test is how to set the proof-stress level. The classical approach consists of measuring the slow crack growth constants, setting-up a proof test diagram taking into account the service stress, and, finally, setting the proof-stress/service-stress ratio so as to get the minimum lifetime required.

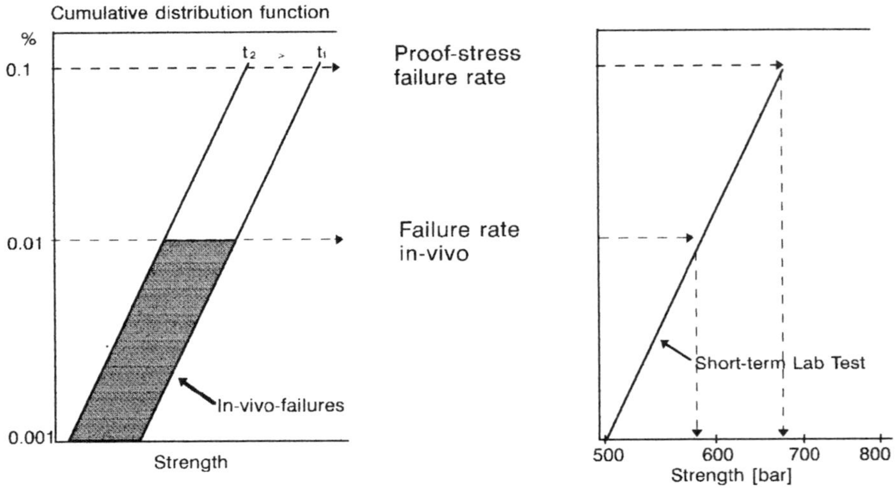

Fig. 3: Determination of the proof-stress level

In the present investigation the proof stress level is not based on the in service stress because this stress is not documented precisely enough. On the other hand, what is very well documented is the in-vivo failure rate encountered in 20 years of clinical experience. It is these data that are used for determining the proof-stress level.

The determination of the proof-stress level from the failure rate is done in the following steps. According to basic life-time calculation procedures there is an interdependence between the sustainable load and the time under loading:

(3) $$\sigma_t^{\,n} \cdot t = \text{const.}$$

Figure 3 shows the distribution of in-vivo failures at loading times t_1 and t2 based on equation (3) and the measured strength distribution function of hip joint heads measured in the proof-test set-up. Knowing the stress distribution function under proof stress conditions at failure rates between 0.01 and 0.01 % and knowing that the in vivo-failure rate is around 0.01 one can determine the corresponding short term loading level. As a consequence, using this loading level as the proof stress level one would eliminate all balls which are prone to in-vivo failure. This is based on the worst case assumption that all in vivo failures are due to weak balls. By using a proof-stress level which corresponds to a failure rate of 0.1 %, i.e. that is 10 times the in-vivo failure rate, one is far on the safe side.

Summary

A proof-test for hip joint heads has been set up which uses a liquid for pressure loading. The load is applied to the ball head only on the surface of the conical bore. The efficiency of the proof-test for ceramic hip joint heads is shown by comparing the resistance to static load (according to ISO 7206-5) of proof-tested and of non proof-tested hip joint heads. The strength distribution of proof-tested hip joint heads turned out to be truncated at the predetermined level. The determination of the proof-stress level for production purposes is based on the known in-vivo failure rate. The proof-test equipment is operational in the production line.

References

[1] Evans, A.G., Wiederhorn, S.M (1974): Proof testing of ceramic materials - an analytical basis for failure prediction. Intern. J. of Fracture 10 (1974) 379-392

[2] Maier, H.R., Krauth, A., Stärk, N., Zeibig, A. (1980): Long Life Ceramic-Metal HIP joints. in: Winter, G.D., Gibbons, D.F. and Plenk, H. [eds]: Biomaterials 1980, Wiley and Sons, p 219-226

[3] Kleer, G., Soltész, U., Siegele, D. (1985): Applicability of the proof-test concept to ceramic hip-joint heads. in: Lee, A.J.C [ed]: Biological and Biomechanical Performance of Biomaterials, Proceedings of the 5th European Conference on Biomaterials, Elsevier, 1986, p 489-494

LIFE TIME PERFORMANCE OF BIOCERAMICS FOR BIOMEDICAL PROSTHESIS APPLICATIONS

Hengchu Cao
Carbomedics Inc., Austin, TX

Abstract: Pyrolytic carbon is an important bioceramic material for medical implant device applications. However, it is well known that ceramic materials are typically brittle and often subject to subcritical crack growth under sustained loading and/or dynamic cyclic loading. Consequently, a prerequisite for the successful application of the ceramics in safety-critical structural components is the long term mechanical reliability. In this study, the application of a pyrolytic carbon in a heart valve prosthesis is explored. Specifically, an experimental procedure was developed for rigorous characterization of the fracture resistance and fatigue crack extension. Experiments were conducted under sustained and cyclic loading in a simulated biological environment. While the material was shown to exhibit brittle fracture, it displayed excellent resistance to subcritical crack growth. The crack growth kinetics in pyrolytic carbon was formulated using a power law equation. Damage tolerance concept based on fracture mechanics was used to develop an engineering design approach for mechanical heart valve prostheses. A *safe-life* index was introduced to assess the design adequacy against subcritical crack growth in pyrolytic carbon. Such a procedure has been applied in the design of prosthetic heart valve components to ensure structural integrity against fracture and fatigue.

1 Introduction

Carbon exists in nature in a variety of different forms, most commonly in the form of graphite, diamond, buckminsterfullerene (C_{60}) and amorphous carbon. A number of pyrolytic carbons also exist having a turbostratic microstructure between graphite and the amorphous phase. Low-temperature isotropic pyrolytic carbon has been found to exhibit excellent wear resistance, mechanical strength and biocompatability. It is the preferred material in cardiovascular prosthesis and also used in hip joint and dental implant.[1-3]

The mechanical performance of pyrolytic carbon material plays an important role in determining the structural integrity of a prosthesis. Pyrolytic carbon has a high mechanical strength and a low Young's modulus, resulting in a large energy of rupture. The failure strain of pyrolytic carbon (Table I) ranges from 1.2 to 1.8%, compared with most ceramics at 0.1 to 0.2%. The modulus is close to that of the corticle bone. However, pyrolytic carbon is a brittle solid and the fracture behavior is similar to a ceramic. For example, the fracture toughness of pyrolytic carbon is about 1 to 2 MPa√m compared with 3 to 10 for advanced ceramics and > 20 for most metal alloys. This low fracture toughness results in a strength sensitive to the presence of crack-like flaws. In addition, many recent experimental studies have indicated that ceramics are frequently susceptible to environmentally enhanced stress corrosion cracking as well cyclic loading induced fatigue crack propagation[4-10]. Since heart valve applications involve extremely large numbers of loading cycles (i.e. heart beats), substantial crack extension may occur in service under subcritical conditions. To prevent structural failure of mechanical heart valve prostheses, fatigue crack extension must be considered in the design.

In light of the recent observation of subcritical crack growth in pyrolytic carbon[5-7], it is important to re-assess the critical fracture and fatigue properties of the material and to evaluate the impact of these observations on the mechanical performance of the prosthetic heart valve. Previous results did not provide accurate quantitative information on the crack extension velocity, and thus were inadequate for design purposes. It is the intent of this investigation to develop rigorous experimental procedures for the characterization of crack growth properties in pyrolytic carbon. Specifically, fracture toughness, flexure strength distribution and subcritical crack growth rate under cyclic and sustained static load are rigorously characterized using an experimental procedure developed at CarboMedics. Life-time prediction based on the safe-life concept is illustrated using exemplary Pyrolite® carbon heart valve components. A critical implementation of damage tolerant design involves the determination of the initial flaw size. Non-destructive inspection and proof test techniques are explored.

2 Experimental Characterization Procedures

2.1 Fracture toughness and crack growth experiments

Fracture and fatigue crack growth experiments were conducted on ASTM compact disk specimens (Fig. 1). For this study, Pyrolite® carbon, a pyrolytic carbon material produced using Carbomedics proprietary processing, was employed. The pyrolytic carbon coating was deposited on a graphite substrate disk in a fluidized bed composed of ceramic particles and the mixture of inert carrier gas, hydrocarbon and silicon-alloying gases. The coating conditions were controlled to deposit a homogeneous isotropic pyrolytic carbon coating similar to those used in making the Carbomedics Prosthetic Heart Valve(CPHV) components. About 8% silicon was added in the Pyrolite® carbon coating. Addition of silicon in the coating results in the formation of SiC particulates in the matrix and enhances wear resistance[11].

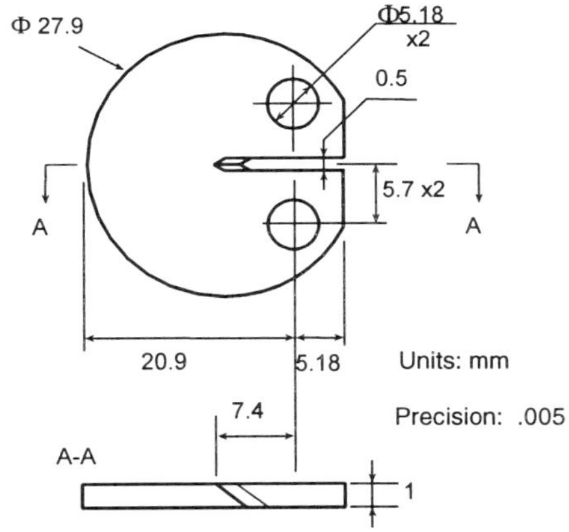

Fig. 1 Geometric dimensions of a typical disk standard compact specimen and the slanted starter notch.

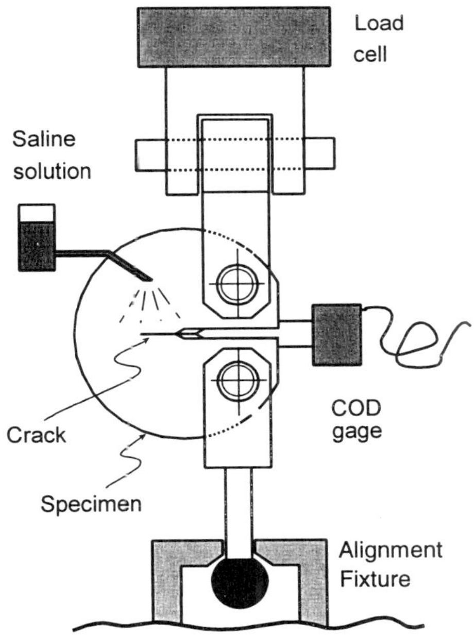

Fig. 2 Schematic showing the compact disk specimen fixtures and the universal alignment jig used to couple specimen to the MTS servo-hydraulic testing system, as well as the Ringer's solution injection system.

As-coated disks contained a graphite substrate covered by the carbon coating. The coating was separated from the substrate, lapped parallel and polished. Loading holes and notches were made using fine grit diamond tools, to the dimensions as shown in Fig. 1. Both surfaces were polished with 1 μm diamond paste to facilitate microscopic crack measurement.

Tests were conducted on a MTS servo-hydraulic machine. The specimens were accommodated to the actuator using a special loading fixture similar to the ASTM standard fixture with one modification(Fig. 2). A universal ball-socket coupling allowed free rotation of the grips to maintain a linear alignment from the specimen to the loading grips. This is important because a slight misalignment could induce nonuniform load distribution across the specimen thickness. The crack mouth opening displacement(COD) was monitored using a high resolution mechanical extensometer, attached to the edge of the specimen. In addition, a traveling microscope with a resolution of 1 micron was mounted on the MTS to monitor the crack length. To facilitate fatigue precracking, a slanted angle notch was used as a crack starter.

The test was controlled with an IBM PC using an integrated control and data-acquisition software. Sinusoidal waveforms with specified frequency, minimum load level and cycling amplitude were supplied through a RS232 interface to the function generator. Load, crack opening displacement and loading point displacement were acquired by the PC and recorded during the test. Crack length was computed from the loading/unloading compliance and verified independently using the traveling microscope. The system can be programmed to conduct a simple ramp R-curve test or a cyclic load fatigue test as well as a constant load stress corrosion cracking test. Tests can be controlled to maintain constant peak load, peak displacement, peak COD or peak stress intensity factor. It can also provide constant ΔK control and decreasing ΔK control to determine fatigue threshold. A typical testing sequence includes: 1) initial fatigue precracking, 2) optical measurement of starting crack length, 3) effective modulus determination through single ramp compliance check, 4) COD controlled fatigue crack growth test, 5) R-curve test by ramping crack opening displacement, 6) stress corrosion cracking test by holding COD constant, 7) decreasing K threshold scheme, 8) constant ΔK control at below threshold level. Not all specimens were tested in the entire sequence whereas some specimens might be tested under several fatigue assays.

After the test, the data was processed to obtain the crack growth rate. Crack growth velocity was calculated using a linear regression numerical scheme over a small crack length increment. Stress intensity factor associated with the crack extension rate was averaged throughout the corresponding cycles. The variation in the stress intensity within each regression data subset was controlled to be no more than 4%.

The stress intensity factor for the DC(T) specimen is given by,

$$K_I = \frac{P\, f\,(a/W)}{BW^{1/2}} \tag{1}$$

where P is the applied load, B and w are the specimen thickness and width respectively, a is the crack length, f is a dimensionless function [12, 13]. Crack length was obtained by measuring the compliance defined on the loading and unloading curves [14],

$$C_m = \frac{d\delta}{dP} \tag{2}$$

and by the following equation,

$$a/W = g(U) \tag{2a}$$

where δ is the crack mouth opening displacement, g is a dimensionless polynomial[12] and U is defined by[12, 13],

$$U = \left(1 + \sqrt{E'BC_m}\right)^{-1}. \tag{3}$$

with E′ being the effective modulus calculated from the initial compliance using optical measurement of initial crack length including the fatigue precrack.

The ASTM standard imposes two size requirements on the specimen dimensions. Pyrolytic carbon has very high yield strength, $\sigma_y > 500$MPa as determined by the indentation technique. In this case, the characteristic dimension is L = 10 μm. Consequently, the size requirement of test specimens can be easily satisfied for any reasonable specimen. In the present study, we choose W = 20 mm and B = 1 mm.

Since the material is used in an environment exposed to a media with high salt concentration, the human blood, the fracture and crack growth experiments were also conducted in a similar environment. For this purpose, a blood-analog was used when testing a group of specimens. A lactated Ringer's solution was supplied to the specimen by a fluid injection system(Fig. 2).

2.2 Residual stress in disk specimens

Presence of residual stress in fracture test specimens can alter the total crack driving force through the release of the residual strain energy as the crack propagates. Residual stress arises in the coating from the deposition process. Intrinsic stress develops at the deposition due to growth feature interaction during nucleation and growth of the crystallites [15]. Additional residual stress results from differential thermal shrinkage during cooling. When the coating is separated from the graphite substrate, removal of the geometrical constraint can result in a partial release of the residual stress in both the coating and the substrate. The residual stress in the monolithic disk specimen is due to the intrinsic stress gradient. The intrinsic stress is related to the coating system and coating deposition condition. Prediction of this stress is complex and beyond the scope of this paper. However, this residual stress can be measured using an experimental procedure[7].

For pyrolytic carbon disk specimens, an array of strain gages were bonded to each side of the disk along a diameter. The gages were connected to a strain conditioning unit. Then the disk was sliced into three sections to allow stress relaxation. At this point, the beam curved due to the release of residual strain. The residual strain is equal to the strain reading in magnitude but opposite in sign. Residual strain in a disk specimen along the tangential direction was obtained as a function of the position relative to the centroid. Residual stress is calculated from the measured strain by Hooke's law,

$$\sigma_R = E\ \epsilon_R \tag{4}$$

where E is the Young's modulus of Pyrolite® carbon.

3 Results

3.1 Fracture resistance curve

The fracture resistance curve for pyrolytic carbon is shown in Fig. 3 as a function of the crack length. It is evident by examining Fig. 3 that the apparent toughness decreases as a function of crack length, and the variation in K_R depends on the crack length, a, instead of crack growth increment Δa. Since crack length indicates the relative location of the crack tip in the disk specimen, this suggests that either the fracture toughness of the material is inhomogeneous or an additional crack growth driving force is not accounted for by the stress intensity calculation. It was found that residual stress present in the disk specimen contributes to crack growth driving force. To verify this, two DC(T) specimens were heat treated at 1700°C for 2 hours for annealing. The fracture resistance curves for the annealed specimens are measured and shown in Fig. 4. It is clear that except at very small Δa, the fracture toughness is a constant, independent of the crack growth or total crack length. The steady-state fracture toughness ranges from 1.4 to 1.8 MPa$\sqrt{m}$, with a mean of 1.6. The apparent R-curve at the initial crack growth is related to subcritical crack growth. Comparing Fig. 4 with Fig. 3, the difference is thus attributed to the residual strain energy release rate. This correction is also used in the subsequent fatigue experiments for unannealed specimens.

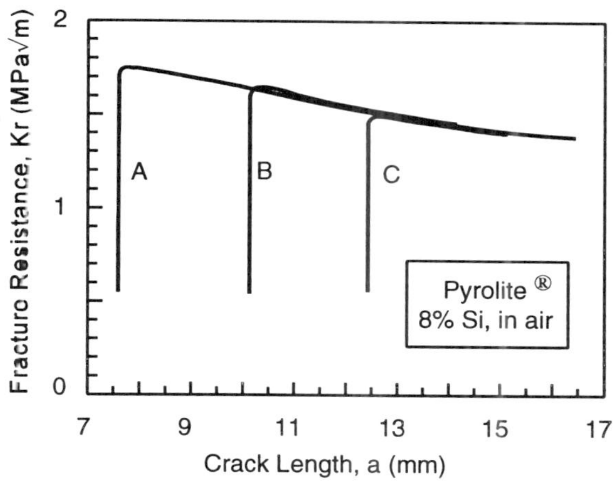

Fig. 3 R-curve measured on as-deposited Pyrolite® carbon showing the fracture resistance dependence on the crack tip location in the specimen.

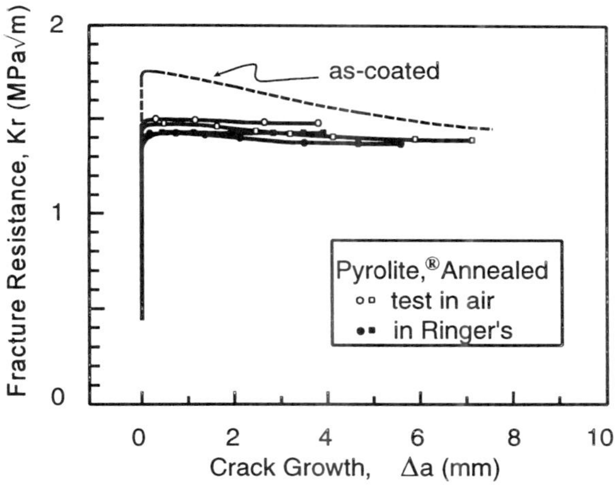

Fig. 4 Effect of heat treatment on the fracture resistance of Pyrolite® carbon.

3.2 Residual stress

Residual stress in the specimen characterized using strain gage technique provided the following stress distribution for the tangential stress component,

$$\sigma_R = \sigma_o \left[1 - 3r^2/b^2\right] \tag{5}$$

where r is the distance from the center of the disk, b the radius of the disk and σ_o represents the stress at the center of the disk with an average of $\sigma_o = -2.2$ MPa. The residual stress is compressive in the middle and tensile at the perimeter.

Pyrolite® carbon specimens were annealed and tested in lactated Ringer's solution. The fracture resistance behavior was similar to those obtained when tested in air. A similar steady-state fracture toughness was also observed.

3.3 Fatigue and stress corrosion crack growth rate

Results for the fatigue crack growth experiment under load ratio R = 0.1 are shown in Fig. 5 for as-deposited Pyrolite® carbon tested in air. In this plot, the stress intensity was calculated from the load only. Consequently, the crack growth rate would depend not only on the crack driving force, K or ΔK, but also on the location of the crack tip due to the residual stress effect. This effect was illustrated with a specimen which was cycled at an initial crack length a_1, and then tested again at the initial crack length a_2. At the same crack growth rate, the applied stress intensity was higher on the first group of loading cycles than on the second. This was attributed to the compressive residual stress present at the specimen center and tensile stress at the edge. When the residual strain energy release rate was considered, the crack growth rate plot was shown to be independent of the crack location, similar to that of the annealed specimens (Fig. 6).

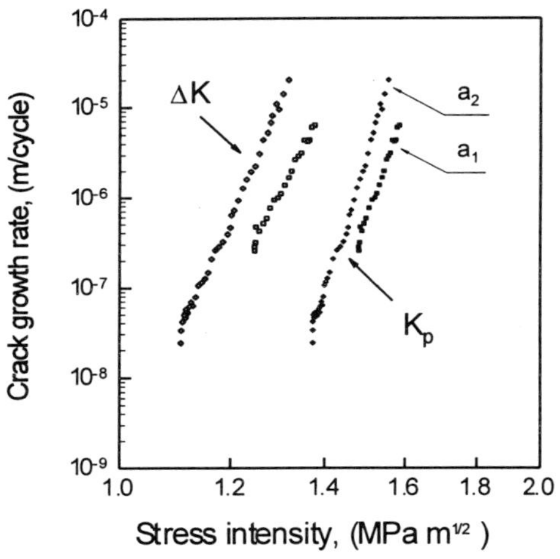

Fig. 5 An example of the crack growth rate as a function of the peak stress intensity factor tested in air, showing the existence of a fatigue threshold as well as the dependence on the crack location in a specimen.

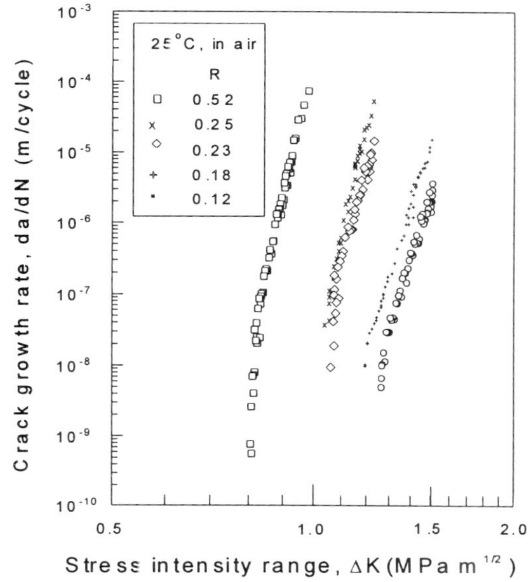

Fig. 6 The effect of cycle load ratio, R, on the measured crack growth rate is demonstrated using the stress intensity factor range ΔK as the crack driving force.

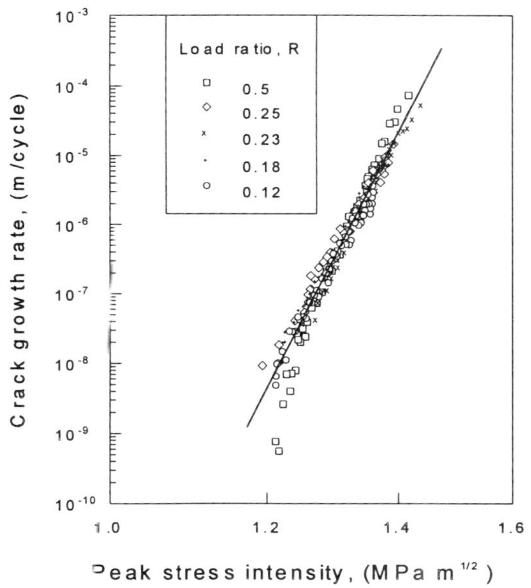

Fig. 7 Crack growth rate as a function of the total stress intensity factor of Pyrolite® carbon tested in air suggests the peak stress intensity as the crack growth driving force, not the stress intensity range.

Comparison of the results obtained on specimens tested in air with those tested in Ringer's solution revealed that the solution has very little effect on the crack growth rate of this material.

To study the effect of the cyclic load ratio on the crack growth rate, cyclic fatigue tests were conducted in air with cycle ratios ranging between R=0.1 and R=0.6. The results, plotted using ΔK and peak K_p as the x-axes (Fig. 6 and 7), indicate that the peak stress intensity, K_p, rather than stress intensity range, ΔK, governs the fatigue crack growth rate. This is in sharp contrast to the fatigue behavior of metals where crack growth is clearly controlled by the stress intensity range[16]. As the load ratio, R, approaches 1 and ΔK approaches zero, the crack growth rate will be controlled by the stress corrosion mechanism. A fracture mechanism transition at cycle ratio between 0.6 and 1 clearly requires further study.

The following empirical equation proves to provide the best fit to all experimentally collected results on Pyrolite® carbon,

$$\frac{da}{dN} = C\left(\frac{K_p}{K_{FC}}\right)^n \tag{6}$$

where $C=10^{-4}$ m/cycle, K_p is the peak stress intensity factor, K_{FC} is the fatigue toughness defined by extrapolating the crack growth rate to da/dN = C, and n is a curve-fitting parameter, known as the fatigue exponent. The constant C has the same dimension as the crack growth rate. The parameters were obtained for each specimen and the mean values are summarized in Table II. It is noted that the K_{FC} defined by this procedure is very close to the plain strain fracture toughness of the material, K_{IC}. If K_{FC} is not available, K_{IC} can be used as a good approximation.

The results for stress corrosion cracking rate are presented in Fig. 8. It is also noted that the testing environment does not have a noticeable effect on the crack growth rate. The da/dt curves obtained in air overlap with those tested in the Ringer's solution. Rate equation similar to Eqn 6 is used to provide the description of the experimental data,

$$\frac{da}{dt} = A\left(\frac{K_I}{K_{SCC}}\right)^\alpha \tag{7}$$

where $A = 10^{-4}$ m/sec. is the proportional constant, α is the stress corrosion susceptibility exponent and K_{SCC} is the stress corrosion toughness defined by extrapolating the crack growth rate to da/dt = A. The parameters in Eqn 7 are summarized in Table II.

Table I Material properties of Pyrolite® carbon and graphite

Material	POCO Graphite	Pyrolite® Carbon
Young's modulus (GPa)	11 ~ 14	28 ~ 32
Poisson's ratio	.15	.21
Density (g/cm^3)	1.8	2.1
Weibull modulus, m	5 ~ 9	12 ~ 18
Weibull constant σ_o (MPa)	85	320
Fracture strength (MPa)	50 ~ 120	350 ~ 500
Failure strain (%)	~ 1	1.2 ~ 1.8
Crystallite size, l_c (nm)	300 ~ 5000 †	30 ~ 50

† Average particulate size for POCO graphite.

Table II Fracture and fatigue properties of Pyrolite® carbon

Fracture toughness, K_{IC} (MPa $\sqrt{m}$)	1.5
Fatigue toughness, K_{FC} (MPa $\sqrt{m}$)	1.45
Fatigue exponent, n	55 ~ 80
Fatigue constant, C (m/cycle)	10^{-4}
Fatigue threshold, K_{TH} (MPa $\sqrt{m}$)	1.2
SCC toughness, K_{SCC} (MPa $\sqrt{m}$)	1.45
SCC susceptibility exponent, α	150 ~ 180
SCC constant, A (m/sec.)	10^{-4}
MOR of leaflet beams, (MPa)	450
MOR of slabs, (MPa)	400

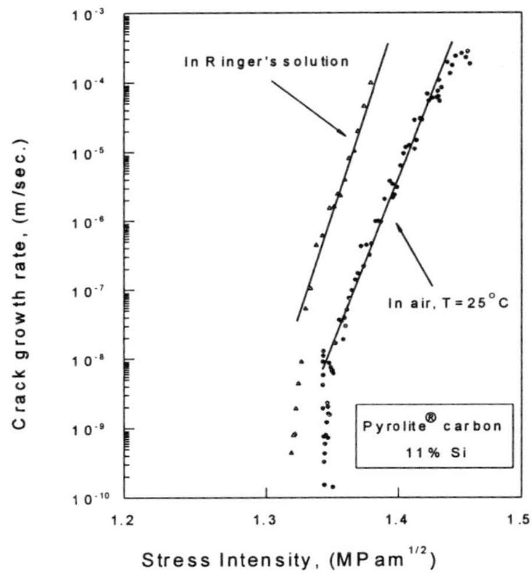

Fig. 8 Subcritical crack growth rate under static load was measured using Pyrolite® carbon specimens in air and Ringer's solution.

4 Damage Tolerant Design Procedure

4.1 Basic Design Concept

The concept of damage tolerant design has been widely used in the design of safety critical structures, such as boilers, high pressure pipelines, nuclear reactors and aircraft [17, 18]. This procedure was further adopted for cardiac valve prostheses [5,19]. Two basic design philosophies exist in the design against catastrophic brittle fracture. The *fail-safe* concept achieves its objective by providing sufficient built-in redundancy in the case of individual component failure, thus permitting subsequent repair or replacement. Since replacement or repair of a dysfunctional valve involves open-heart surgery and thus imposes additional threat to the survival of the patient, this approach is not desirable. On the other hand, the *safe-life* design concept assures the safe operation of the structure within its design life. This design philosophy requires that the flaw present in the structure be well characterized and smaller than an allowable size. Thus the designer must have detailed knowledge of the complete loading conditions during the life of the structure, and the effect of such loading on the material properties.

4.2 Life-time prediction

Life time prediction should be performed on every component based on the most highly stressed region, thus the primary mode of failure. Since a bi-leaflet mechanical heart valve prosthesis consists of three components of complex geometry, identification of the primary failure mode is typically conducted by computational mechanics. For example, for a mitral valve, one of the most likely failure paths in the leaflet is a corner crack located half way across the B-datum on the in-flow side where it is subjected to greatest stress (Fig. 9). The pertinent stress intensity factor solution for this scenario can be approximated by considering a corner crack in a plate with thickness, h, subject to bending [14].

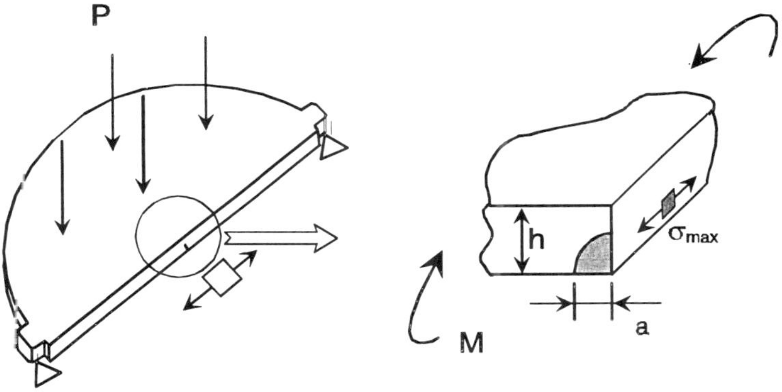

$$K_I = \sigma_{max}\sqrt{\pi a}\,Q(a/h)$$

Fig. 9 Schematic for damage tolerant analysis showing a potential crack on the B-datum of a leaflet where the applied stress is the greatest.

Initial flaw size, a_i, is defined by inspection and proof-test, and the critical flaw size, a_c, is defined by design stress as,

$$\left.\begin{aligned} a_i &= \frac{1}{\pi}\left(\frac{K_{IC}^{U}}{Q\sigma_p}\right)^2 \\ a_c &= \frac{1}{\pi}\left(\frac{K_{IC}^{L}}{Q\sigma_a}\right)^2 \end{aligned}\right\} \tag{8}$$

where σ_p is the proof-test stress, σ_a is the worst-case service stress, K^U_{IC} and K^L_{IC} are the upper and lower bound fracture toughness, respectively. Integration of Eqn. 6 gives,

$$N = \int_{a_i}^{a_c} \frac{1}{C}\left(\frac{K_{IC}}{Q(a,h)\sigma\sqrt{\pi a}}\right)^n da \tag{9}$$

For a small circular corner crack, a<<h, the Q(a,h) function becomes a constant, $Q(a,h) = 2.4/\pi$, and the integration of Eqn. 9 can be carried out explicitly, leading to,

$$N = \frac{2a_i}{(n-2)C}\left(\frac{K_{IC}}{Q\sigma\sqrt{\pi a_i}}\right)^n \left[1-\left(\frac{a_i}{a_c}\right)^{n/2-1}\right] \tag{10}$$

Noting that a_i is typically much smaller than a_c, the above equation can be simplified as,

$$N \approx \frac{2a_i}{(n-2)C}\left(\frac{K_{IC}}{Q\sigma\sqrt{\pi a_i}}\right)^n \tag{11}$$

Since n is large for any brittle material, the lifetime prediction using Eqn. 11 is very sensitive to the actual values used in the service stress σ and the initial crack length a_i. Noting that

$$\frac{K_{IC}}{Q\sqrt{\pi a_i}} = \sigma_1 \;(\equiv \sigma_p) \tag{12}$$

represents the minimum initial strength of the component, Eqn. 11 can be rearranged to obtain,

$$\sigma_N = \lambda_N \sigma_1 \tag{13}$$

where σ_N is the safe-life cyclic stress that a component with initial strength σ_1 can sustain continuously for a design life time of N cycles, and λ_N is referred to as the *fatigue degradation factor*, given by,

$$\lambda_N = \left(\frac{2a_i}{(n\text{-}2)NC} \right)^{1/n} \tag{14}$$

A similar procedure can be applied to the rate equation for stress corrosion cracking. The total life time can be obtained by integrating Eqn. 7 for a small initial crack of a_i, leading to,

$$t_f \approx \frac{a_i}{(\alpha/2\text{-}1)A} \left(\frac{K_C}{Q\sigma\sqrt{\pi a_i}} \right)^{\alpha} \tag{15}$$

Rearranging Eqn. 28, a *stress corrosion degradation factor* can be defined by,

$$\lambda_t \equiv \frac{\sigma_t}{\sigma_1} = \left(\frac{a_i}{(\alpha/2\text{-}1)t_f A} \right)^{1/\alpha} \tag{16}$$

where σ_t is the safe-life static stress that a component with initial strength σ_1 can sustain continuously for a design life time of t_f (in seconds).

As an example, the safe life-time for a CPHV leaflet can be computed from the following parameters: σ_p =167 MPa, $a_i < 21\mu$m, σ_a = 50 MPa, C=10^{-4} m/cycle, Q=2.4/π [1]. In this case, we obtain the following life time for the failure mode of concern: N = $5.5\text{x}10^{38}$ cycles, ie. 10^{30} years, and t_f= 6x 10^{71} sec. i.e., 10^{64} years.

4.3 The Safe-Life Index Analysis

In engineering design practice, a safety factor is usually defined by the ratio of the characteristic strength of the material over the design stress. This factor reflects the potential to tolerate design variability and the reliability of the structure. To design against subcritical crack growth under cyclic fatigue and stress corrosion cracking, the *safe-life index* is defined by,

$$\mathbf{S} = \lambda_t \lambda_N \sigma_1 / \sigma_a \tag{17}$$

where σ_a is the design stress and σ_1 the minimum initial strength warranted by proof-test or other means. It is noted that σ_1 is not a material constant, but depends on the initial flaw size detected on the actual component of concern.

A fatigue safe-life index of 1.2 to 2 is recommended for designing safety-critical structural component with brittle materials. It must be noted that this safety margin was imposed in addition to the use of the worst-case service load, initial flaw size as well as conservative material properties. Taking the example given above, the safe-life index is computed to be **S** = 1.8 for a design life of N = $4x10^9$ cycles. This indeed indicates a conservative design with respect to the fatigue induced subcritical crack growth. This procedure must be performed for all critical components and at the critical failure locations in the design process.

4.4 Flaw detection and proof test

As indicated in the above section, determination of the initial flaw size in a structural component is a very important step to perform the damage tolerant analysis. There are a range of crack detection techniques available which enable a component to be examined without being damaged. A reliable non-destructive evaluation technique must be able to detect cracks below the critical size at critical locations where a crack is most likely to extend during service. Each technique has its limitation in the size of the flaw that can be detected.

Visual inspection makes use of the light contrast formed by defects on a smooth surface under reflected light. The operation is conducted using the naked-eye aided with low power magnifying optics. It is efficient and effective in detecting surface features and defects. The resolution is determined by operator training, surface finish of the parts and the magification power of the optics.

Surface cracks are a major concern for ceramic materials. Dye-penetrant testing is a method that involves the application of a liquid penetrant over the surface area to be inspected. After the penetrant is allowed to be fully absorbed, it is wiped off the surface before a powdered 'developer' is applied. When an ultraviolet dye is used, crack-like defects and porosity will produce a fluorescent contrast under the illumination of an ultraviolet light. Under a 40x inspection microscope, flaws of dimension about 50μm can be readily detected. Rigorous training and experience of the operator can further improve the detection limit.

Proof-test is an effective way to reject defective parts. It can be used as a NDE technique provided that it does not propagate a crack in the accepted component. After passing a proof test stress, σ_p, the component will not fail in service as long as the service stress is always below σ_p. Since subcritical crack growth can result in delayed failure, an additional safety margin must be allowed in the design stress. This margin is governed by the damage tolerant design analysis.

Proof test, optical-visual inspection, dye-penetrant examination and function testing are combined to ensure flaws present in a component smaller than the critical size. Throughout the manufacturing process, each and every orifice and leaflet are inspected with the aid of optical microscopes and dye-penetrant. To detect flaws smaller than those detectable by optical methods, a mechanical proof-test is conducted on each component and again on each orifice-leaflets subassembly to reject assembled valves not meeting critical flaw criterion. These redundant procedures are used to ensure that the flaw size in any carbon component is absolutely below the designed "tolerable" flaw size. Finally, the Instructions For Use (IFU) emphasize the importance of proper handling to avoid the introduction of surface flaws during the valve replacement operation.

5 Conclusion

Pyrolite® carbon is biocompatible, wear-resistant, durable and resistant to stress corrosion and fatigue cracking. The fracture toughness measured using ASTM standard compact disk specimen ranges from 1.4 to 1.8 MPa$\sqrt{m}$ for Pyrolite® carbon. The material exhibits excellent resistance to fatigue and stress corrosion cracking. The crack growth kinetics in pyrolytic carbon was formulated using a power law equation. The measured fatigue exponent is greater than 55, compared with 2 to 4 for most metals. The SCC susceptibility exponent is even larger, $\alpha = 180$, compared with glass and ceramic($\alpha = 7$- 15).

Damage tolerace concept based on fracture mechanics was used to develop an engineering design approach for mechanical heart valve prostheses. A *safe-life* index was introduced to assess the design adequacy against subcritical crack growth in pyrolytic carbon. Such a procedure has been applied in the design of prosthetic heart valve components to ensure structural integrity against fracture and fatigue. In cardiovascular applications, over 25 years of successful clinical experience with prosthetic heart valves made of Carbomedics Pyrolite® carbon attests to the effectiveness of this design approach and to the resistance of this material to fatigue. It is shown that damage tolerant design combined with quality control in manufacturing and effective non-destructive flaw detection can ensure safe life-times for prosthetic heart valves.

References

1. J.K. Ryder and H.C. Cao, "Structural integrity assessment of heart valve prostheses: a damage tolerant analysis of the Carbomedics Prosthetic heart Valve (CPHV)," J. Heart Valve Dis., in press.
2. F. Schoen, "Carbon in heart valve prostheses: Foundations and clinical performance," Biocompatible Polymers, Metals, and Composites, Chap. 11, pp.239-261, Technomic, Lancaster, PA, (1983).
3. A.D. Haubold, H.S. Shim and J.C. Bokros, "Carbon in medical devices," Biocompatibility of clinical implant materials, Vol. 2, ed. D.G. Williams, CRC press, Boca Raton, FL, (1981).
4. G. Singh and L. Ma, "Long-life cyclic fatigue of pyrolytic carbon," Bioceramics, 6, pp. 211-215, (1993).
5. R.O. Ritchie, R.H. Dauskardt, W. Yu and A.M. Brendzel, "Cyclic fatigue-crack propagation, stress corrosion and fracture-toughness behavior in pyrolytic carbon-coated graphite for prosthetic heart valve applications," J. Biomed. Mat. Res. (1990).

6. C.B. Gilpin, A.D. Haubold and J.L. Ely, "Fatigue crack growth and fracture of pyrolytic carbon composites,", Bioceramics, 6, pp.217-224, (1993).
7. H.C. Cao, "Fracture resistance and cyclic fatigue behavior of pyrolytic carbon", 6th Int. Sym. on Ceram. in Med., PA, (1993).
8. L. Ewart and S. Suresh, "Crack propagation in ceramics under cyclic loads," J. Mat. Sci., 22, pp.1173-92, (1987).
9. R.H. Dauskardt, R.O. Ritchie and B.N. Cox, "Fatigue of advanced materials: Part II," Advanced Materials & Processes, 8, pp.30-35, (1993).
10. S. Wiederhorn, "Environmental stress corrosion cracking of glass," NBS Report 10-565, Apr. 1971.
11. H.S. Shim and F. J. Schoen, "The wear resistance of pure and silicon-alloyed isotropic carbons," Biomat. Med. Dev., Art. Org., 2, 103-118, (1974).
12. ASTM Standard E399-90, "Standard Test Method for Plane -Strain Fracture Toughness of Metallic Materials," Annual Book of ASTM Standards, American Society for Testing and Materials, Philadelphia, PA, 1992, section 3, vol. 3.01, pp506-536.
13. ASTM Standard E647-91, "Standard Test Method for Measurement of Fatigue Crack Growth Rates," Annual Book of ASTM Standards, American Society for Testing and Materials, Philadelphia, PA, 1992, section 3, vol. 3.01, pp674-701.
14. R.W. Hertzburg, Deformation and fracture mechanics of engineering materials, John Wiley & Sons, (1983).
15. R.W. Hoffman, "The mechanical properties of thin condensed films," Physics of thin films, Chap. 3, pp.211-273, (1966).
16. P.C. Paris and F. Erdogan, "A critical analysis of crack propagation laws," Trans. ASME, J. Bas. Engng, 85, pp528-34, (1963).
17. ASME Boiler and Pressure Vessel Code (1977).
18. Airplane Damage Tolerance Requirements, USAF Military Specification MIL -A-83444. (1974).
19. R.O. Ritchie and P. Lubock, "Fatigue life estimation procedures for endurance of a cardiac valve prostheses: stress life and damage tolerace analysis," ASME J. Biomech. Engr, 108, pp.153-160, (1986).

WEAR OF ENAMEL AGAINST GLASS-CERAMIC, PORCELAIN, AND AMALGAM

L.A. George and F.C. Eichmiller
American Dental Association
Paffenbarger Research Center
Bldg. 224 Room A153
NIST
Gaithersburg, MD 20899

The wear of human enamel opposing certain dental materials is a problem in restorative dentistry. A wear study was carried out on enamel that was in contact with three dental restorative materials, a β-quartz glass-ceramic, a fused porcelain (Ceramco, Inc., Burlington, NJ), and a dispersed phase amalgam (Dispersalloy™, Johnson & Johnson, Skillman, NJ). Twenty-seven human teeth tooled to form enamel cones were randomly assigned to the three restorative materials. Three groups of 9 enamel cones each were abraded against sample disks of each material. The test was carried out on a two-body, unidirectional, rotary wear testing machine. The wear of the enamel was evaluated by measurement of the volume of enamel lost during testing. The average wear volumes lost (cm^3) for the enamel were: glass-ceramic, 5.0×10^{-5} ($s=2.1 \times 10^{-5}$); porcelain, 5.2×10^{-4} ($s=5.4 \times 10^{-4}$); amalgam, 3.5×10^{-5} ($s=1.8 \times 10^{-5}$). Comparison of the mean worn volume of enamel lost for the three materials, by the Tukey method of multiple comparisons with a significance level of 5% , indicated no significant difference between only the enamel worn against glass-ceramic and against amalgam. The results indicate that the β-quartz glass-ceramic material would produce less wear to opposing enamel than traditional fused porcelains. The wear of enamel opposing glass-ceramic should be equivalent to that resulting from contact with amalgam.

INTRODUCTION

The surface hardness of materials is often associated with the abrasion of the materials or of the tooth structure that these materials oppose. Abrasion is an intricate process involving the interaction of many factors; therefore, its use alone as a predictor of wear is generally not valid. Wear has been studied by using a variety of methods, including clinical testing, simulated service measurement, wear testing machines, and evaluation of other mechanical properties. These methods of testing measure only the volume of material lost, they do not determine the mechanism of the wear (1).

A two-body wear test has been used to rank the wear resistance of restorative materials and their effect on opposing tooth structure (2). The abrasive particles in a two-body wear test are attached to one of the material surfaces such as the glass and lucite particles on the surface of porcelain. In a three-body wear test the particles are loose and slide between the two rubbing surfaces (3).

Glass-ceramic inserts (Beta-Quartz Glass-Ceramic Inserts, Lee Pharmaceuticals, South El Monte, CA) are being used in composite posterior restorations. These inserts, which increase the filler loading of the restorations by acting as a large "megafiller" particle within the restorations, are capable of displacing 50-75% of the composite material required to fill the cavity defects. As a result of the decrease in the amount of composite required, volumetric polymerization shrinkage is much lower, and the overall restoration takes on more of the advantageous properties of the insert and maintains dimensional stability. These inserts are often opposing tooth structure; therefore, it is important to determine if they will abrade the opposing structure.

The purpose of this investigation was to evaluate the wear of human enamel in contact with three dental restorative materials, the β-quartz glass-ceramic, a fused porcelain (Ceramco, Inc., Burlington, NJ), and a dispersed-phase amalgam (Dispersalloy^{TM}, Johnson & Johnson, Skillman, NJ) by using a two-body, unidirectional, rotary wear testing machine.

METHODS AND MATERIALS

Enamel Preparation

Twenty-seven extracted human teeth were tooled to form enamel cones. The most conical cusp of each tooth was chosen, and the remaining cusp was ground flat with a high-speed dental lathe and carbide cut-off discs (Austenal

Dental, Chicago, IL). The root portion of each tooth was removed with a low speed diamond saw (Minitom, Struers, Denmark). The teeth were mounted with epoxy (Plastic Steel 5 minute Putty (SF) 10240, Devcon, Danvers, MA) on flat head screws in a special jig that centered and aligned the axis of the screw with the vertical axis of the remaining cusp. Each cusp was precision ground to a 60 degree cone taper by using a mounted high speed handpiece and rotating the cusp. The tip of each cone was ground perpendicular to the cone-rod axis, leaving a flat surface 0.2-0.6 mm in diameter. An initial measurement of the cone diameter was made prior to testing on a light microscope (Wild Heerbrugg, Switzerland) with a measuring ocular. The teeth were stored in distilled water through the preparation procedure to avoid dehydration of the samples.

Glass-Ceramic Disks

Nine glass disks, 3 mm thick and 20 mm in diameter, were prepared with a lost wax casting process by centrifugally casting in a modified Dicor casting machine (Dentsply, York, PA). Following divestment, the glass disks were cerammed by using a modified Dicor ceramming oven (Dentsply, York, PA) following a predetermined heat treatment schedule to produce a glass-ceramic whose primary phase was β-quartz solid solution. The top surface of the disk was finished with a sequence of SiC paper of grit 240, 320, 400, 600 on a water cooled sanding table (Handimet Grinder, Buehler Ltd., Evanston, IL). Following finishing with the SiC paper, the disks were further finished with 1000, 1200, 2400, and 4000 grit SiC paper on an automatic polisher (Planopol, Struers, Denmark). The final finishing of the disks was carried out by polishing the disks on a sequence of 6, 3, 1, 1/4 micron charged diamond wheels on a polishing unit (Planopol-3, Struers, Denmark).

Porcelain Disks

Nine wax patterns, 1 mm thick and 20 mm diameter, were cast with a base metal alloy (Rexillium III, Jeneric Pentron Inc., Wallingford, CT). The cast metal disks were removed from the sprue portion of the casting and were sanded with 240 grit SiC paper. The disks were then treated in a mixture of water and a cleaner (Sparkleen, Fisher Scientific, Pittsburgh, PA) and placed in an ultrasonic cleaner for 3 minutes. The surfaces of the disks were then sandblasted with aluminum oxide to obtain a matte finish, rinsed in ethanol in an ultrasonic cleaner for three minutes and air dried. The surfaces of the disks were then oxidized by using a two-stage heat treatment in a porcelain oven (Centurion VPC, J.M. Ney Co., Bloomfield, CT). Two coats of opaque porcelain followed by three coats of body porcelain (Ceramco porcelain, Ceramco, Inc., Burlington, NJ) were applied according to the manufacturer's recommendations. The disks were machined flat and polished with an automatic polisher (Planopol, Struers, Denmark) in a multiple specimen holder. The disks were polished with SiC paper, 60, 120, 220, 320, 500, 800, 1000, 1200, 2400, and 4000 grit. Polishing of the disks was completed by the use of diamond-charged wheels, 6, 3, 1, and 1/4 micron, on a polishing unit.

Amalgam Disks

Nine amalgam disks, 1 mm thick and 20 mm diameter, were prepared from a dispersed-phase amalgam (Dispersalloy, Johnson and Johnson, Skillman, NJ) by using a three-part, hardened-steel mold and a hydraulic press. The disks were consolidated under a pressure of 1500 pounds and stored overnight in an oven (37 °C) prior to finishing. The disks were finished by using a series of SiC paper, 1000, 1200, 2400, and 4000 grit, on a low speed automatic polisher.

Specimen Mounting

Each disk was positioned in a metal cup filled with dental tray material (Coe Tray Plastic, GC America Inc., Chicago, IL) with the top of the disks parallel to the top of the metal cup. The metal cups were attached to a two-body, unidirectional, rotary wear testing machine (2) that produced unidirectional sliding as recommended in previous research (4). This machine has three rotating cups for mounting the disk-shaped samples and has loaded armatures for accurately positioning the enamel cones on the sample surface. All enamel cones were subjected to a vertical load of 815 g including the weight of the rotor arm. The cones were abraded for 20 cycles, each cycle being comprised of 100 disk revolutions for a time of 75 minutes. The load and time values were chosen because the combination produced measurable wear on samples run in a pilot study without wear penetration into the underlying dentin.

A 3^2 complete factorial experiment was carried out. The two factors involved were as follows: (1) rotor at three levels - rotor 1, rotor 2, and rotor 3, and (2) material at three levels -glass-ceramic, porcelain, and amalgam, for a total of 9 treatments. The experiment was replicated three times and blocked by day for a total of twenty-seven runs in three days. The stationary enamel cones were set perpendicular to a rotating disk of glass-ceramic, porcelain, or amalgam in a medium of distilled water at 37 °C.

The volume of enamel worn from each cone was calculated from the cone diameter and height before and after testing. The formula used to calculate the worn volume was that used by Palmer (5) and illustrated in the figure below (5).

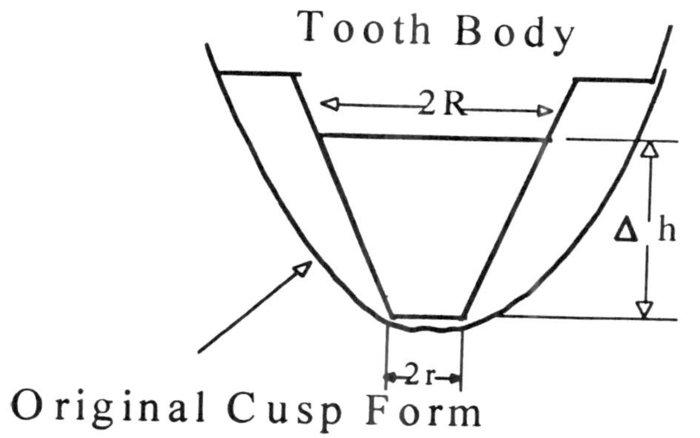

Worn Volume $= \Pi(\Delta h)^2 (r^2 + rR + R^2)/3$

Δ h: change in height

r: initial radius

R: final radius

The worn-enamel cone volumes were adjusted for differences in the radius of the track worn into the disks by using the Locke approach (6) to standardize the effective distance traveled by each enamel cone against the opposing disk. Statistical analysis of the data was carried out with SAS (7) computer software package (SAS Institute, Inc., 1987). ANOVA was carried out to compare the means of the wear volume opposing the three materials.

RESULTS

The average wear volumes lost (cm^3) for the enamel were: glass-ceramic, 5.0×10^{-5} ($s=2.1 \times 10^{-5}$); porcelain, 5.2×10^{-4} ($s=5.4 \times 10^{-4}$); amalgam, 3.5×10^{-5} ($s=1.8 \times 10^{-5}$). Comparison of the mean worn volume of enamel lost for the three materials, by the Tukey method of multiple comparisons with a significance level of 5%, indicated no significant difference between the enamel worn against glass-ceramic and against amalgam. While enamel worn against porcelain was significantly different from the rest.

DISCUSSION

The inhomogeneous microstructure of fused porcelain leads to an inherent roughness in the hard surface of dental restorations. Fused porcelain can be glazed or polished to a very smooth surface, however, natural degradation in the oral environment results in an extremely abrasive surface when the individual glass and lucite particles from the fused glass matrix become exposed. Such results have been seen clinically with destruction of opposing tooth structure and restorations (8, 9,10,11,12).

A glass-ceramic based on stuffed beta-quartz has been developed for use in inserts in composite restorations (13,14). This material is being further developed for use as a single tooth restorative material that would provide for inlays, onlays, and crowns. These restorations would possess the desirable properties of porcelain, be moderately easy to fabricate, would not shrink during firing, and would match the color and translucency of tooth structure without external coloration.

CONCLUSION

These results indicate that the microcrystalline nature of the glass-ceramic may impart superior wear properties. The glass-ceramic, which had hardness values nearly twice the hardness of the porcelain (15), did not show higher wear of opposing enamel by wear volume. In fact, the glass ceramic exhibited wear of opposing enamel equivalent to dental amalgam. The initial smooth surface of the porcelain disks was worn away quickly due to the inhomogeneous microstructure; the wear resulted in the exposure of individual glass and lucite particles as the soft glass matrix was abraded. The surface of the glass-ceramic samples remained smooth and showed no evidence of surface degradation due to abrasion against enamel. These results justify the use of this type of glass-ceramic material in its present application as an insert in composite posterior restorations and encourage studying its use as posterior prostheses such as inlays, onlays, and crowns.

ACKNOWLEDGMENT

This investigation was supported, in part, by USPHS Grant P50 DEO9322 to the American Dental Association Health Foundation from the National Institutes of Health - National Institute of Dental Research and is part of the dental research program conducted by the National Institute of Standards and Technology in cooperation with the American Dental Association Health Foundation.

Certain commercial materials and equipment are identified in this paper to specify the experimental procedure. In no instance does such identification imply recommendation or endorsement by the National Institute of Standards and Technology or the ADA Health Foundation or that the material or equipment identified is necessarily the best available for the purpose.

REFERENCES

1. Craig R G, ed. (1985). *Restorative Dental Materials*, St. Louis: The C.V. Mosby Company, p. 99.

2. McKinney JE (1982). Apparatus for measuring wear of dental restorative materials. Wear 76:337-347.

3. Phillips RW(1991). *Science of Dental Materials*, Philadelphia: W.B. Saunders Company, p. 559.

4. Powell JM, Phillips RW, Norman RD (1975). In vitro wear response of composite resin, amalgam and enamel. *J Dent Res* 54:1183-1195.

5. Palmer DS, Barco MT, Pelleu GB, McKinney JE (1991). Wear of human enamel against a commercial castable ceramic restorative material. *J Prosthet Dent* 65(2):192-195.

6. Locke JCG (1976). A method of testing the wear properties of porcelain and human tooth enamel [Thesis]. Indianapolis: Indiana University.

7. SAS Institute Inc. (1988). *SAS/STAT Users Guide*, Release 6.03, Cary, NC: SAS Institute Inc., pp. 597-598 and 946-947.

8. DeLong R, Sasik C, Pintado MR, Douglas WH (1989). The wear of enamel when opposed by ceramic systems. *Dent Mater* 5:266-271.

9. Krejci I (1991). Wear of Cerec and other restorative materials. In: Mormann WH, editor. State of the art of the CEREC-method. Proceedings of the International Symposium on Computer Restorations, 1991 May 3-4, Regensdorf-Zurich (Switzerland): *Quintessence*, 245-251.

10. Craig RG, Powers JM (1976). Wear of dental tissues and materials. *Int Dent J* 26(2):121-133.

11. Seghi RR, Rosenstiel SF, Bauer P (1991). Abrasion of human enamel by different dental ceramics in vitro. *J Dent Res* 70(3):221-225.

12. Hayes D, Seghi RR, Rosenstiel SF (1993). Abrasion rates of human enamel opposing various dental ceramics (abstract 888). *J Dent Res* 72(SI):214.

13. Bowen RL (1991). Microcrystalline insert for megafilled composite dental restorations, U.S. Patent 5,057,018, October 15.

14. George LG, Eichmiller FC, Bowen RL (1992). An intrinsically colored microcrystalline glass-ceramic for use in dental restorations. *Ceram Bull* 71(7):1073-1076.

15. Winnert LA, Bowen RL, Eichmiller FC (1991). Heat treatment effects on hardness of a beta-quartz glass-ceramic composition (abstract 1746). *J Dent Res* 70:484.

1995 Corporate Award Winners

1995 Corporate Technical Achievement Award:

Transelco Div., Ferro Corp.

X5000 Dielectric for Multilayer Capacitors

X5000 Dielectric for Multilayer Capacitors is an advanced, state-of-the-art, preformulated composition that features the highest dielectric constant available for X7R-type multilayer ceramic capacitor (MLCC) materials.

MLCCs are extensively utilized in today's sophisticated circuitry for performing functions such as filtering, tuning, coupling/decoupling, isolation, and storage of electrical energy. MLCCs are found in increasing quantities in areas such as telecommunications, data processing, consumer electronics, transportation, etc.

Technological advances such as supercomputers, on-dash automotive command centers, e-mail, and even high-definition television have been helped along by the continued evolution of miniaturization and the continuous increase in circuit board and substrate densities for electronic subsystems. MLCCs have, by necessity, followed along in this miniaturization process, meeting increasing demands for smaller and more powerful capacitors.

MLCC manufacturers have been successfully striving to increase capacitance density by reducing physical size while increasing volumetric efficiency. Volumetric efficiency can only be enhanced by increasing the relative dielectric constant (K´) and/or by decreasing the layer thickness of the dielectric.

The Ferro X5000 dielectric provides manufacturers with the optimum volumetric efficiency in the most popular capacitor type, i.e., X7R. By using microstructural engineering techniques in the development of X5000, Transelco researchers were able to produce a dielectric constant >5000, which approaches the theoretical limits for fine-grained barium titanate ceramics.

X5000 can be utilized to advance the state of the art by taking advantage of the increased dielectric

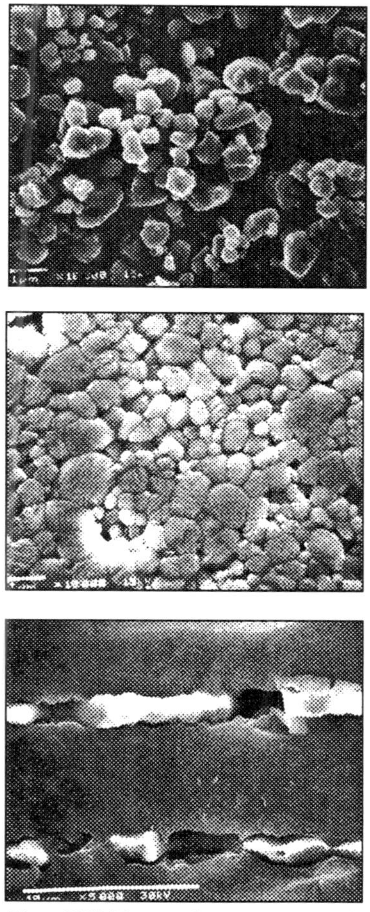

Top: *X5000 powder as received.* ***Middle:*** *X5000 thermally etched microstructure.* ***Bottom:*** *XJ4000 thin layer cross-section (dielectric thickness ≅7 µm).*

constant or by reducing layer thickness by as much as 50%. MLCCs now can be produced with fewer layers per unit, resulting in a substantial reduction in the cost of precious-metal internal electrodes. X5000 uses specially doped, chemically derived, titanate material to obtain optimal dielectric constant, which, according to the literature, occurs in the 0.7–1.0 µm grain-size area.

Transelco's synthesis technique yields a uniform material. Heterogeneous grain size in X5000 is controlled to an average of 0.8 mm (see etched photomicrograph). This results in an increased number of grains per dielectric layer thickness and greater resistance to electrical failure, compared with standard, coarser grained materials of the same dielectric thickness.

Since the X5000 composition was patented in 1992, a derivative product, XJ4000, has been developed especially for the Japanese MLCC market and for those applications requiring improved DC bias and aging characteristics. This product has a somewhat lower dielectric constant, 4000, but has other features of significant interest to MLCC manufacturers, e.g., a grain size of ~0.5 mm.

1995 Corporate Technical Acheivement Award:

Norton Advanced Ceramics

Cerbec Silicon Nitride Bearing Components

Cerbec® Silicon Nitride Bearing Components are produced by Norton Advanced Ceramics, a subsidiary of Saint Gobain/Norton Industrial Ceramics Corp., East Granby, Conn. The components are densified by hot isostatic pressing, utilizing a glass encapsulation technology licensed from ACC of Robertsfors, Sweden. Rolling elements in the form of balls, rollers, and raceways are fabricated in diameters from 1 to >100 mm. These components are used in rolling element bearings to improve performance in a variety of applications.

Norton's NBD200 bearing material offers significant improvement in many high-performance bearing applications. The material's unique combination of high hardness and toughness allow direct replacement for certain conventional steel components. Improvements in bearing life, increased speed, reduced heat generation, and improved corrosion resistance are some of the performance gains providing increased value to high-speed bearing applications.

Additionally, improved performance under lubrication starvation conditions is a system benefit. Because of silicon nitride's high modulus of elasticity, improved system stiffness is often exhibited as well. Cerbec silicon nitride's material properties and reliability have demonstrated significant bearing system performance by improving quality and reducing downtime in many applications.

These performance increases have clearly demonstrated added value for many bearing users and has justified the higher cost over the conventional technology. Major markets for the product are industrial (high-speed machine tool spindles) and medical (high-speed dental drills). Development in aerospace applications also is a focus.

The present technology is patterned after conventional pressed and sintered ceramic components to minimize cost. Powder compacts are formed using spray-dried raw materials. These compacts are encapsulated in glass and hot isostatically pressed at pressures near 2000 bar and temperatures >1800°C. After the encapsulant is removed, the net-shape ball blank is ground and lapped to its final dimensions, using technology similar to that used in metal ball finishing. Cost reduction is a key focus of Norton to expand market penetration.

Part of the success of the product has been Norton's ability to engineer the microstructure to meet the needs of the bearing applications. Bearing components are among the most precise man-made articles. Sphericity of the silicon nitride ball products is typically >1 min. (0.025 mm), allowing the fabrication of bearings with little vibrational noise. This has been made possible by creating a fine and uniform microstructure. This microstructure also allows excellent wear resistance and rolling contact fatigue resistance.

Cerbec Silicon Nitride Bearing Components manufactured by Norton Advanced Ceramics.

General Issues

COMMERCIALIZATION OF THE IN SITU VITRIFICATION TECHNOLOGY

James E. Hansen
Geosafe Corporation
2950 George Washington Way
Richland, WA 99352
(509) 375-0710

John S. Tixier
Pacific Northwest National Laboratory (PNNL)
Battelle Boulevard
Richland, WA 99352
(509) 376-8732

ABSTRACT

In situ vitrification (ISV) is a thermal treatment technology for the environmental remediation of soil sites contaminated with hazardous (organic and inorganic) compounds and/or radioactive materials, along with buried waste and debris. The process involves application of electrical energy to the soil which is progressively converted to a molten mass. Inorganic contaminants are incorporated into the pool while organics are destroyed by the advancing thermal front. Once the molten mass cools, it solidifies into a highly durable glassy to crystalline monolith which is estimated to retain its integrity for geologic periods.

Geosafe Corporation is the sole commercial provider of ISV technical and remediation services through exclusive license rights to the patented technology. Geosafe initiated commercialization of ISV in 1988 following the invention and development of ISV by Battelle for the U.S. Department of Energy in the early to mid-1980's. Significant challenges, barriers, and setbacks to commercialization have been encountered and overcome in bringing the new technology to the marketplace. A review of the ISV technology development and commercialization processes is presented.

INTRODUCTION

The purpose of this paper is to present a description of the commercialization process experienced in bringing a capital-intensive technology with fairly limited development experience to an arena wherein the numerous forces, both outside and self-imposed, acting against successful commercialization are very difficult to foresee and are often beyond the direct control of those trying to direct the process. As opposed to some of the other highly successful technology development and commercialization examples presented in this Symposium, the ISV commercialization process encountered a number of challenging difficulties that had to be overcome. Therefore, the current successful position of the company and the commercialization of ISV in the U.S. and internationally is all the more meaningful in light of the challenges, barriers, and setbacks to commercialization that have been encountered and overcome in bringing the new technology to the marketplace.

Since ISV is probably not what one might consider a typical ceramic technology, the following summary description of the ISV technology is presented to assist the reader in understanding the type of technology being commercialized. The ISV technology has been recognized by environmental technologists and regulators as one possessing strong potential for significant contribution to the national problem of hazardous, radioactive, and mixed waste site remed iation and waste treatment. The technology involves the in situ electric melting of earthen materials for two main purposes: 1) thermal destruction/ removal of hazardous organic and inorganic compounds, and 2) permanent immobilization of hazardous and radioactive heavy metals within a high integrity vitrified product form. In addition to beneficial site remediation applications, several possible useful purposes for ISVare also envisioned including conversion of waste materials into recycled products, and production of high integrity vitrified masses for various civil engineering applications.

Invention and development of the ISV technology grew out of the joule-heated glass melting technology that was adapted to waste management applications by PNNL starting in the early 1970s. The technology differs from standard melter technology in that the process is conducted without containment other than the earthen media itself. To initiate the in situ melting of soil, an array of electrodes (usually four graphite rods) is placed a few inches into the media to be treated, and a graphite and glass frit "starter" material is placed in paths between the electrodes. The starter material acts as an electric heating element when power is applied through the electrodes. As the starter material heats up, the energy is transferred to the surrounding soil which also begins to heat up and melt, eventually consuming the starter material. Once molten, the earthen media (e.g., natural soil) becomes electrically conductive and acts as the process heating element. Joule heating occurs within the molten mass. Heat is conducted outward into adjacent unmelted media, causing it to melt, thereby enlarging the size of the melt. This process continues until the desired treatment volume/mass is attained. Single melts of up to 1,400 tons have been generated by ISV. Expansive sites are treated by repeating the process in a batch sequence. Adjacent melts fuse together to form a single contiguous monolith.

Within the melt, temperatures in the range of 1,500 to 1,800 degrees Celsius are experienced with most silica-based media (e.g., soil, sludge, sediment, mine tailings). Gases generated during the process are confined by a sheet metal hood that is placed over the site being treated.. Gases are drawn from the hood and processed through an off-gas treatment system, consisting of a series of wet scrubbing and High-Efficiency Particulate Air (HEPA) filtering components, before being exhausted to the atmosphere. Additional unit operations (e.g., carbon adsorption, thermal oxidation) can be incorporated into the off-gas treatment train, if necessary, based on site specific contaminants and conditions. Secondary waste materials generated by the off-gas treatment can be recycled back to the ground for treatment by a subsequent melt. Figure 1 illustrates the overall ISV process and equipment.

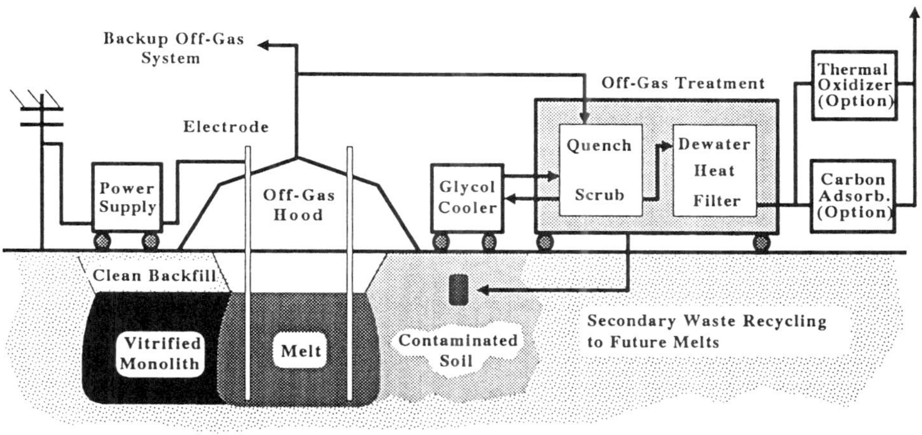

Fig. 1. Overall ISV Process and Equipment Arrangement

The high temperature of the melt causes the pyrolytic destruction and vapor-phase removal of organic contaminants present in the treatment volume. Most metal oxides become part of the melt and are chemically incorporated within the glass and crystalline structure that is produced upon cooling. The process results in a large volume reduction due to the elimination of void volume within the particulate media (e.g., 20 to 50% for most soils). The vitrified product possesses outstanding physical, weathering, leaching, and biotoxicity properties; it is physically and chemically analogous to igneous rock formations such as basalt or obsidian. The ISV process and its attendant vitrified product may be the most technically preferred and cost effective environmental restoration technology alternative in cases where 1) a combination of organic and heavy metal, or hazardous and radioactive, contaminants must be treated; 2) on-site and in situ treatment is highly desired; 3) the site contains nonhomogeneous soil conditions and/or debris; 4) maximum treatment effectiveness and permanence are required; and/or 5) high volume reduction is desired.

TECHNOLOGY DEVELOPMENT AND COMMERCIALIZATION CHRONOLOGIES

The chronology of the ISV technology development and commercialization efforts can best be understood in relation to two key timelines: technology development and technology commercialization. The timelines are presented separately below; however, it should be noted that they are closely related and somewhat interdependent. It should also be noted that two separate technology development efforts, DOE's and Geosafe's, have been pursued since Geosafe's establishment. The DOE program is mentioned in the discussion below since, although separate from Geosafe's efforts, it had major influence of the perception of ISV in the marketplace.

Technology Development Chronology

The ISV technology development timeline is depicted in Figure 2, in which key milestones are identified relative to the technical development, test and demonstration, and application of the technology. Invention and proof-of-principle testing of ISV occurred in 1980. The following two years focused on initial process exploration and small-scale test and development work, all of which continued to indicate the promise of the technology. There was a strong desire by DOE for rapid technology scale-up; thus, starting in 1983, the program moved away from laboratory process and technology development efforts to a pilot-scale (i.e., capable of producing 10- to 50-ton melts) field equipment design and demonstration effort. The initial scope of DOE application interest at that time was for TRU-contaminated soil. Pilot-scale equipment was designed and fabricated, and a major milestone was achieved in 1983 by the successful completion of a demonstration test on a staged TRU-contaminated soil site.

This success resulted in rapid movement of the program toward design, fabrication, and demonstration of a large-scale ISV equipment system capable of treating 100 ton/day of contaminated soil, and designed to be trailer mounted for over-the-road mobility. Some equipment development work was also performed in the areas of electrode design (for extended life) and electronic depth monitoring equipment (for determining the depth of melt in real-time). The multimillion dollar large-scale (full-scale) ISV process system was completed in 1984, an amazingly short four years after inception of the technology concept. Four large-scale Operational Acceptance Tests (OATs) were performed in uncontaminated soil during the following 1-1/2 years. The equipment was then used in 1986 to conduct the Large-Scale Radioactive Test (LSRT) which treated a portion of a TRU-contaminated drain field within Hanford's 200 Area. This test was largely successful; however, some equipment difficulties were encountered, and the depth attained was less than desired. It was learned that the depth had been limited by the presence of a cobble rock layer that served as a thermal barrier to melt growth. This finding was very important to the future development of ISV. During the next four years, other the technology development efforts were expanded as the large-scale equipment sat idle during this time.

In the meantime, the promise of the technology became attractive to other DOE sites resulting in the establishment of several separate development and test programs for various types of applications, including the following:

- radioactive liquid waste disposal seepage basins at ORNL (initiated in 1985),
- buried waste at INEL (initiated in 1987), and
- underground radioactive waste storage tanks at Hanford, ORNL, and other DOE sites (initiated in 1989).

Pilot-scale demonstrations, using both chemical simulants of the radioactive target materials and radioactive tracers, were conducted in staged settings. Since the pilot-scale ISV equipment was housed in one semi-trailer, the tests were conducted at the sites identified above. In addition, in 1984, PNNL began laboratory-scale tests and demonstrations on many types of

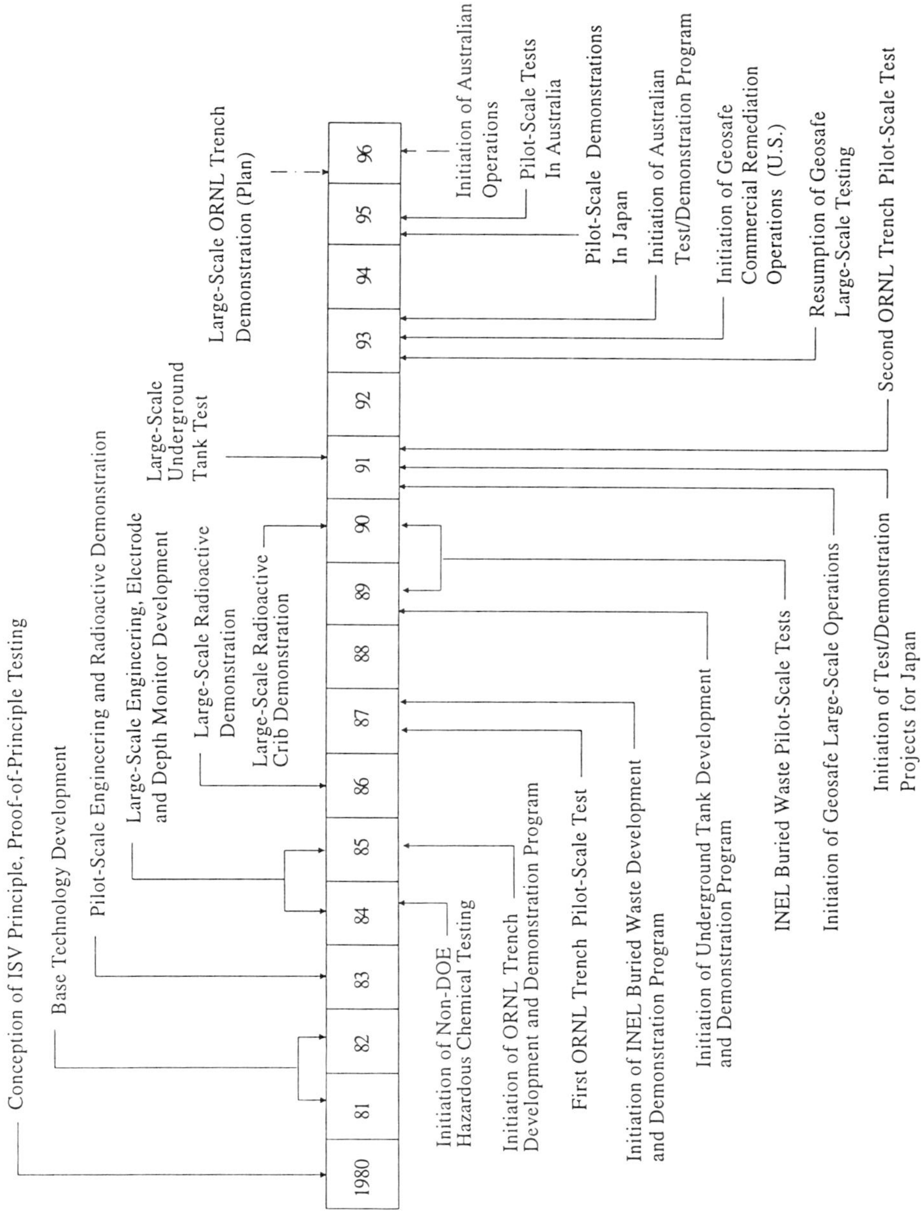

Fig. 2. ISV Technology Development Milestone Timeline

hazardous chemical wastes for non-DOE clients. The technology was found to be applicable for the effective treatment of a broad range of earthen media and contaminant types. In 1989 Geosafe procured and initiated testing of a large-scale ISV machine (system) for its commercial use.

In 1991, the technology was applied to another large-scale demonstration at Hanford involving a radioactive liquid waste disposal seepage crib (wooden timber type). While quite successful in most regards, this demonstration, like the LSRT, came up short in the area of depth attainment. These two large-scale tests had identified the inability to minimize melt width as needed to allow attainment of desired melt depth. Many ideas for resolving the problem were identified by PNNL staff; however, they were unable to obtain the funding needed to resolve this problem. This was the apparent nature of the DOE program at that time (i.e., demonstrations were readily funded, basic science and R&D was not).

In addition to the depth limitation results, in the late 1980's, several ISV field tests experienced notable difficulties. First, during pilot-scale testing of ISV on buried wastes at INEL, PNNL found that the flow of vapors and gases through the melt was sometimes sufficient to disrupt and splatter the molten soil, making process control difficult and ultimately causing damage to the equipment. (At that time,a high-temperature ceramic fabric was being pursued for use as the off-gas hood; difficulty was encountered in maintaining the integrity of the material upon prolonged or direct exposure to molten soil.) Second, during very aggressive operational acceptance testing of its new large-scale equipment, Geosafe explored new limits of application. These efforts included buried drums containing water, and resulted in a melt displacement and splattering event that severely damaged the ceramic fabric off-gas hood that was being employed. (Following this second event, a decision was made to discontinue testing and use of the ceramic fabric hood material.) Third, during a large-scale PNNL demonstration test on a 22,000 liter (6000 gal) underground tank in a simulated setting, a vent tube failed midway through the melt, resulting in a thermal excursion, displacement of molten soil, and damage to the off-gas steel hood shell and support structure. The vent pipe, and other mitigating tools, had been deployed in the underground tank vitrification test following successful results in a similar pilot-scale test. None of these three events were injurious to human health; however, they clearly revealed an area of technical difficulty that had not been a concern for simple contaminated soil applications: that being the need for process control and equipment design features capable of handling the volumes of gases that could be rapidly generated during such applications.

As a result of the difficulties experienced in these field tests, instead of pursuing the scientific and engineering resolution of the issues, DOE decided to curtail the technology development efforts. DOE's newly created Office of Technology Development (OTD) took the position that the ISV technology was sufficiently developed for unconfined contaminated soil type applications, required little or no further development work, and should be considered as an "available technology" for DOE remediation needs. Since, in its view, any further development needs should be conducted on a site specific basis, OTD also terminated its ISV Integrated Program which had been established to investigate the underlying technical principles that needed to be understood to successfully implement ISV for the various applications. DOE took the position that further development of "advanced" ISV applications (e.g., buried wastes and underground tanks) would wait until ISV became an established technology for contaminated soil applications. The DOE buried waste and underground tank ISV programs essentially terminated at that point; however, several relatively small test and development projects related to contaminated soil applications at ORNL, Savannah River Site, and Hanford were maintained.

On the commercial side, Geosafe proceeded to investigate the cause of its testing event and, in collaboration with researchers at PNNL, developed a significant advancement in understanding of the generation and disposition of gases and vapors during ISV processing. Geosafe established new application limits and process control standards, and designed and constructed a new sheet metal off-gas hood capable of withstanding higher off-gas temperatures as well as exposure to occasional splatter from molten soil. Geosafe resumed testing of its new equipment early in 1993 and initiated its first commercial remediation project during the Summer of 1993. Whereas the early stages of the DOE ISV development program focused on proof-of-concept, equipment design and development, and test, development and demonstrations (and, when possible, scientific explanation and understanding of ISV phenomena) of interest to DOE sites and applications, Geosafe's efforts have focused on process dynamics, process control, equipment refinement, expansion of application types, and developing the know-how required to make the technology commercially successful on a production basis. These efforts have resulted in a greatly increased ISV knowledge base being established since the first Geosafe field melt in 1991.

Geosafe's first commercial project was at the Parsons Chemical Superfund Site in Grand Ledge, MI, with melt operations beginning in the summer of 1993. That project involved the treatment of 4,800 tons of soil contaminated with pesticides, mercury, and low levels of dioxin and arsenic. The project involved some site and soil conditions that had never been tested at large-scale before; and Geosafe had to respond with several adaptations to the equipment and process. The challenges were successfully met, and a highly successful EPA Superfund Innovative Technology Evaluation (SITE) Demonstration Program demonstration and evaluation was performed and a full report produced by EPA. This site ultimately required 9 melt settings and almost a year to complete.

Geosafe then performed a National Toxic Substances Control Act (TSCA) Demonstration at a private site in Spokane, WA, with melt operations beginning in the fall of 1994. That project involved the treatment of 3,100 tons of soil and debris (ruptured drums, concrete, asphalt) contaminated with PCBs to a maximum concentration of 17,800 ppm. That project was very successful with no significant process or equipment challenges being encountered. Five melt settings were required over the course of about 3 months. After all the process data and site sampling and post-test characterization was completed, Geosafe was awarded a National TSCA Operating Permit for processing PCBs throughout the U.S.

Geosafe's third commercial remediation project was performed at the Wasatch Chemical Superfund Site in Salt Lake City, where about 6,000 tons of soil and debris contaminated with dioxin, pentachlorophenol, herbicides, pesticides, and a variety of other organics were processed. Melt operations began in winter 1995 and the 37 melt settings were completed in just under one year. This project marked the first ever onsite and in situ thermal treatment of dioxins. The project was highly successful and met all requirements imposed by EPA and the State of Utah.

Limited technology development work is currently being sponsored by DOE, with only two projects currently active, both funded by ORNL Environmental Restoration. The first project involves a full-scale treatability study on an inactive radioactive liquid waste seepage trench at ORNL where researchers plan to use 3 melt settings to process about 1000 tons of soil contaminated with about 40 Ci of radioactivity. The treatability study is the culmination of 10 years of ISV research and development collaboration between PNNL and ORNL, including two highly successful pilot-scale ISV tests at ORNL. Because of possible volatilization of radioactive compounds, special filters are being employed in the off-gas stream that are designed to reduce generation of secondary waste and excessive worker exposure to radioactivity. An new, state-of-the-art hood system has been designed and constructed for the ORNL application. The treatability study will allow researchers to evaluate and refine various aspects of ISV processing conditions and equipment on a soil site with a reasonable level of radioactive contamination before ORNL proceeds with the technology for remediation of other sites contaminated with much higher levels of radioactivity. A second ISV project is investigating possible ISV application to buried wastes at ORNL using a series of laboratory-scale experiments at PNNL. Possible modifications to the baseline ISV technology being investigated on this project include pretreatment of a buried waste site by dynamic compaction or processing the buried waste site by applying ISV upward from the bottom of the site, or a possible combination of the two. The goal of this project is to mitigate possible pressurization excursions that can be experienced when ISV is applied to intact buried containers.

Geosafe's technology development efforts are currently focusing on work for pending non-DOE applications within the U.S., and for overseas clients including contaminated soil, LLW, and waste treatment applications in Japan, and a major buried waste application involving hazardous and TRU contaminants at the Maralinga Test Range in South Australia.

ISV Commercialization Chronology

The ISV commercialization timeline is depicted in Figure 3. PNNL scientists and engineers invented the technology in 1980. A patent application was filed in 1981, and a very fundamental patent was granted to DOE in 1983. Similar patents were granted by major industrialized countries throughout the world. Patents for a variety of technology improvements have been granted over the years, although as with any technology under continuing development, some of the patents are obsolete. In addition to pursuing additional patents, much of the current operating know-how obtained by Geosafe during the recent successful commercial remed iations is treated as proprietary information by the company.

DOE granted a license to Battelle Memorial Institute (operator of PNNL) in 1986. Battelle initiated marketplace promotion and regulatory awareness efforts on a privately funded basis. These efforts resulted in the selection of ISV for remediation use at three U.S. sites: 1) the Pristine Superfund (Remedial) Site in Cincinnati, OH, 2) the Griener Lagoon Superfund

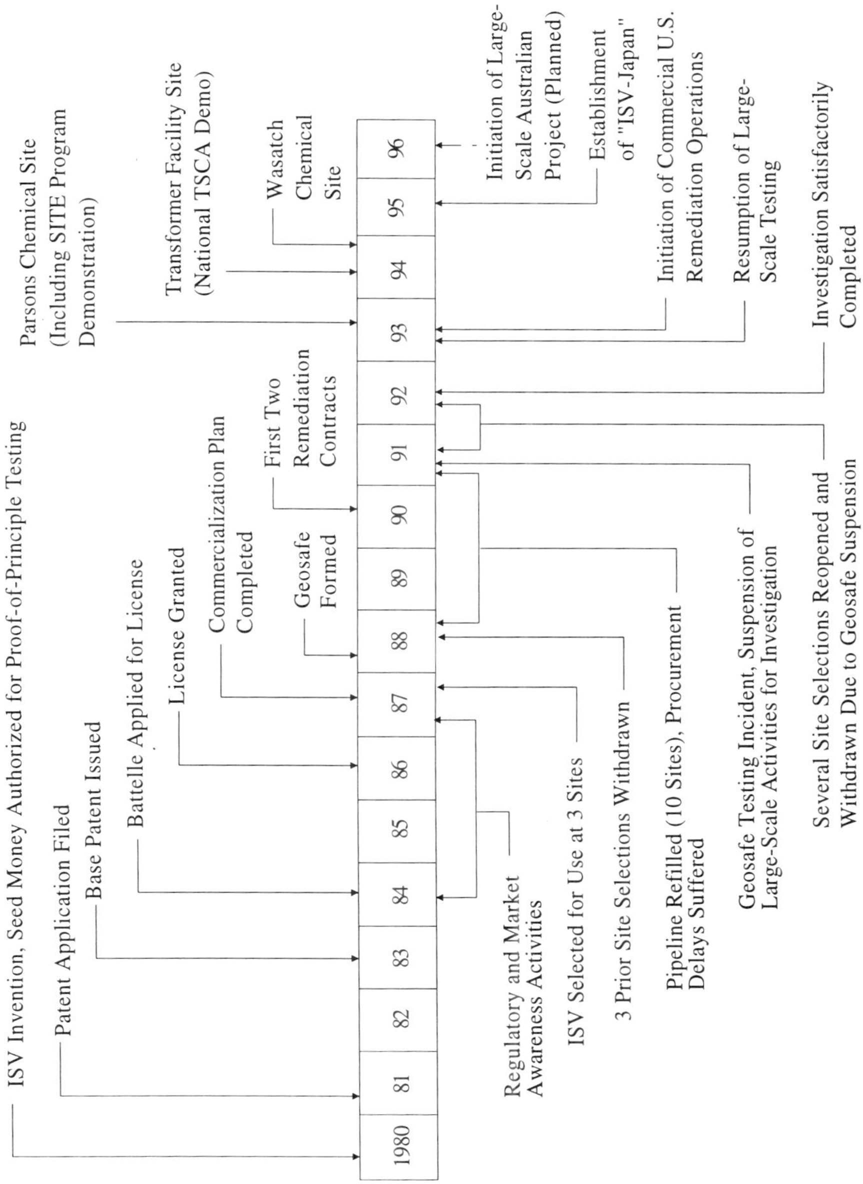

Fig. 3. ISV Commercialization Milestone Timeline

(Removal) Site also in Ohio, and 3) a private RCRA closure site in North Carolina. Battelle defined a strategic commercialization plan that involved creation of Geosafe Corporation in 1988 to fulfill the needs of these selections through an exclusive sublicense. Geosafe initiated procurement of its first large-scale ISV equipment system later that year.

The Company suffered a major setback within one year after its formation as all three of the site selections were withdrawn. The Pristine Site Record of Decision (ROD) was reopened after a surprising turn of events. EPA had issued this ROD and many others in response to Congress' mandate (SARA/1986) to show site selection preference to innovative technologies that involved treatment, reduced toxicity, mobility, and volume, and could be performed on-site and in situ. That mandate essentially specified technologies like ISV for many difficult sites. In response, EPA issued many innovative technology RODs without first having performed treatability tests on the selected technologies on actual wastes from the sites. Congress' Office of Technology Assessment (OTA), in one of its twice yearly critiques of EPA, took EPA seriously to task for having done this. In addition, OTA questioned the wisdom of EPA in making the ISV selection merely because a competitor of Geosafe's had made claims (unsubstantiated) that "ISV doesn't work." In Geosafe's opinion, OTA's challenge to EPA resulted in significant damage to the advancement of innovative technologies, in direct conflict with Congress' mandate.

The Griener Lagoon site selection disappeared in a similarly unanticipated manner. EPA added lime to the site to make the wet soil easier to handle during staging of soil in preparation for ISV treatment. After the site was ready to treat, EPA performed sampling and analysis to confirm the PCB content of the soil. To their surprise, the PCBs had apparently disappeared. Similar results were noted at other PCB sites. This finding was termed EPA's "lime discovery". The ISV site selection was removed while the discovery was further investigated. Some years later it was determined that the addition of lime to the sites did not really destroy the PCBs, rather it may have removed some by volatilization due to the heat generated from the addition of lime, and the presence of lime was found to mask the PCBs making them difficult to detect and measure by the conventional analytical methods then employed.

The private RCRA site selection was lost when Geosafe determined it was not possible to negotiate an acceptable contract with the site owner.

Geosafe was then faced with reestablishing a customer market for its ISV test and remediation services. A significant number of treatability tests were performed in support of Superfund RI/FS work. Within a few months the State of Michigan decided to fund an ISV demonstration at one of its sites. In collaboration with the State, it was determined that EPA would fund the demonstration at the Parsons Chemical Superfund Site under the existing ERCS contract. Geosafe was told to be ready to mobilize to the site within two months. This timing was unfortunate in that EPA administrators shortly thereafter determined that the ERCS contracting provisions should not be used for such projects, wherein a subcontractor (Geosafe) would be doing most of the work. They proceeded to use a competitive contracting approach that stretched the procurement period from the original two months, by a factor of 11 times, to 22 months. Geosafe finally obtained that contract, and another with a private client, during the Fall of 1990.

During these 22 months, Geosafe's marketing efforts yielded significant fruit, with the addition of 10 site (preferred remedy) selections for the ISV technology. These selections involved Superfund RODs and other comparable decision statements by DOD and private parties. At this point Geosafe was vigorously looking for an investment partner to fund the equipment and working capital that would be required to support this anticipated level of demand.

In 1990, Geosafe hired and trained a full operating crew for its new large-scale system. This crew would perform two large-scale melts as part of its training prior to mobilizing to the first project. The first melt went well. The second employed particularly aggressive operating conditions and resulted in the incident reported above in the technology chronology section. Geosafe determined that the incident was significant enough that large-scale commercial operations should be put on hold (suspended) while the cause was investigated. It took about one and one half years for Geosafe to fully evaluate the cause and to determine a recovery plan. An additional half year was consumed designing and fabricating a new sheet metal hood. Geosafe resumed testing early in 1993, and commercial operations were initiated shortly thereafter.

During the suspension of large-scale operations, a number of the site selections understandably went away. Two RODs were reopened and revised due to the uncertainty as to how long the suspension would last. Other "multiple choice" selections which allowed alternatives to ISV resulted in decisions against ISV during this time. Two DOD demonstration selections were withdrawn because of budgetary constraints and Geosafe's inability to meet desired mobilization schedules. At the end of the suspension, three of the original selections remained, and two more had been added. Geosafe then initiated commercial operations in 1993 with five intended selections for site remediation by ISV.

Geosafe currently has completed remediation of three of these sites. The attainment of commercial operations status has been helpful in restoring awareness by the marketplace and regulatory community that ISV has returned after the suspension. At this time ISV is being seriously considered for many possible private and government sites, both in the U.S and abroad.

CHALLENGES IN THE ISV COMMERCIALIZATION PROCESS

Significant challenges, barriers, and setbacks to commercialization of ISV have been encountered and overcome in bringing the new technology to the marketplace. The lessons learned in this process are numerous and ongoing. In this section, a description of the various potentially difficult issues associated with commercialization of a technology like ISV is given. The issues are grouped into three areas: technical, financial and contractual, and public relations. Finally, a summary of some of the key lessons learned during the commercialization of ISV is given.

Technical Issues

At the time of Geosafe's inception, the prior ISV test record had led to the belief that the ISV technology was very capable and fully ready for commercial application to soil remediation projects. The technology had been developed and demonstrated at full-scale in an unusually brief time. It was to be realized later, as the technology was tested under more aggressive conditions and for much more complex applications, that the basic underlying technical principles governing the technology had not yet been fully explored nor were they adequately understood, primarily in two areas: 1) melt shape control (which impacted attainable treatment depth), and 2) generation and disposition of gases and vapors during processing. Whereas there had been ample funding for equipment and demonstration projects, funding for basic research on underlying scientific principles was not forthcoming from DOE. EPA also declined to fund experimental and test work because "ISV was DOE's technology".

These factors, in addition to the fact that no two sites are exactly the same (and most are actually very different), required Geosafe to undertake additional efforts in order to understand and offer a viable technology. It also required Geosafe to establish a relatively tight "applications envelope" in which the underlying principles were understood for applications within the envelope. Other applications would require further testing and development. In this sense, then, Geosafe considered the initial state of technology readiness as a type of barrier (or limitation) to full attainment of the technology's commercial potential.

Financial and Contractual Issues

Volatility of Site Selections: It may take many years for a site to move through the process from initial discovery and characterization to final remediation. For purposes of predicting firm demand on which to base company decisions, the technology-linked remediation vendor may look at contracts in hand, which is the firmest indicator, or possibly at official site selections specifying use of the vendor's technology. Since ISV is a sole source technology, Geosafe assumed that site selections specifying the use of ISV were pretty firm. Part of the learned reality was that site selections are not inviolable, and it is common for them to be reopened, overturned, or otherwise made highly uncertain. It was also found that typical remediation contracts allow for the client to terminate the contract without cause, further adding to the difficulty of making credible projections.

Lack of Understanding Regarding Small Commercial Business Needs: Innovative technologies are mostly developed and offered by small business concerns with limited capital resources. Geosafe believes that the needs of such companies are not well understood by the Government organizations attempting to deal with them. For example, EPA's SITE Demonstration Program requires the technology vendor to fund its own demonstration, while the SITE Program will fund data acquisition and evaluation efforts. Such large-scale demonstrations can cost many hundreds of thousands of dollars; small innovative technology companies cannot handle that level of "marketing" cost. The alternatives exist for the developer to find a client to fund the demonstration, or for the Government to provide a demonstration site. In the search for an ISV demonstration site, EPA's SITE Program formally requested DOE to consider providing a site where a demonstration could be performed. For unknown reasons, a suitable site could not be found from the hundreds possessed by DOE. Largely because of this situation, the SITE Program evaluation of Geosafe's ISV technology occurred many years after the technology was accepted into the SITE Program; and even then, the site employed was not the ideal site to satisfy the objectives of the SITE Program.

In similar manner, innovative technology companies are faced with the need to fund large-scale equipment without the existence of firm market demand. The companies are also expected to maintain staff during periods of contracting or regulatory delays of various types. The developer typically is expected to withstand these costs without assistance. The developer may also be expected to provide special low pricing as an incentive to obtain a demonstration opportunity. Such costs can be staggering to a small technology innovative company, and should be recognized by those organizations charged with encouraging the development and implementation of innovative technologies. Such costs also mitigate against profitability, which makes it difficult for companies to raise additional investment capital.

Government Contracting Inconsistencies: The difficulty posed to contractors by the Government contracting process was epitomized by Geosafe's experience wherein a two month procurement stretched into 22 months. The location of the procuring office, and the personnel involved, changed several times during that time. Geosafe found there to be a significant disconnect between the operations side of many Government client organizations, and the legal/contracting side. The standard wisdom is for vendors to work both sides of the house at the same time. However, our experience indicated that the legal/contracting side would not usually get functionally involved with a project until it was time to negotiate a contract. This situation not only caused problems for Geosafe, but for the operations side of the client organization itself. It should be noted that at the time of this experience (1990), many of the procurement practices for remediation projects were in the fairly early stages of development, and were subject to many changes. Geosafe has not noticed the extent of such problems when dealing with private clients; however, as a general rule, it does seem to be more difficult in terms of time and effort to negotiate a contract than to perform the technical work associated with project development.

Geosafe also experienced a case wherein the Government client was always right, so to speak, when it came to interpreting contract provisions. The Government held firm to its "promise" that there would be no change orders during the fixed price contract, regardless of whether or not they were justified. Geosafe was obligated to accept the cost of unforeseen technical challenges and site discoveries during this project. Even extending the duration of the contract extracted a significant additional financial toll. There exists an obvious dichotomy within the Government wherein, on the one hand they are charged with supporting the introduction of innovative technologies, and on the other hand they are charged with being as hard as possible in the area of contract enforcement. Typical development stage innovative technology companies are seldom positioned to withstand such costly "support" from the Government.

Difficult Capital Market: At the time of Geosafe's incorporation, the enthusiasm of investors for new environmental technology companies had just peaked and was on its way down. It is general knowledge today that raising significant capital (e.g., millions of dollars) for environmental ventures is very difficult until the company has passed through the proverbial "valley of death". An occasional exception may be successful at raising significant capital if it can be packaged and sold before any downside difficulties unfold. This situation poses a significant challenge to small venture companies since the environmental remediation marketplace, with its relatively large dollar volume projects, and the long procurement and payable times, typically requires very costly equipment and large working capital reserves. Geosafe has been fortunate to have a committed and experienced parent company (Battelle) that has enabled the firm to survive such challenges of commercialization.

Public Relations

Difficulty Establishing Current Awareness: It is a major challenge for an innovative technology company to establish and maintain a current awareness of its technology within the marketplace, including the potential client, regulatory, and environmental engineering segments. Many different organizations (e.g., EPA, DOD, DOE) attempt to assemble and publish standardized data on available technologies on a periodic basis. However, in most cases, the information is seriously out of date (usually at least two years) before it finally gets published and into the hands of users. This problem is particularly important to innovative technology vendors wherein the technology status is usually changing quite rapidly. Significant improvements have been made in this area by EPA's VISITT database and SITE Demonstration Program. However, even now, Geosafe routinely encounters decision makers that are using outdated information on ISV. In an effort to maintain a better current awareness level within the remediation marketplace, Geosafe initiated a periodic publication called the In Situ Vitrification Technology Update. This newsletter has received very good reviews and is obviously much more timely than Government published information. The challenge remains to get the information into the hands of the appropriate decision makers.

Actions of Detractors: Geosafe considers the loss of the Pristine ROD largely to be the result of the competitor's claims against ISV that contributed to OTA's harsh criticism of EPA for the ISV selection. The competitor's claims have since been shown to be unfounded; however, the damage was done. Other actions by competitors and other uninformed persons resulted in the spread of erroneous information regarding ISV. Geosafe has had to counter many myths, including the following beliefs: 1) that all vapors formed during ISV moved into adjacent soil (the myth of "vapor retreat"), 2) that ISV only worked in sandy soils, 3) that ISV would not work in wet soils, 4) that ISV is extremely energy intensive, and 5) that ISV is extremely costly. Geosafe has had to expend significant resources combatting the damage such myths cause in the marketplace and regulatory community.

Loss of Momentum: The series of unexpected testing incidents, and Geosafe's temporary suspension of large-scale operations, had a very negative impact on the momentum and enthusiasm that had been established for the technology. The message went out within the EPA community that ISV was no longer available. Nearly the same effect occurred within the DOE community, as ISV went from being hailed as one of Admiral Watkins' "flagship technologies" and a well funded segment of DOE's technology development program, to a technology that was almost put on standby until it was demonstrated as a commercial success on contaminated soil applications. DOE's disappointment even resulted in "bad press" from the technology's previous number one supporter. This loss of momentum has been and continues to be a major challenge to Geosafe, as it tries to reestablish market awareness of the current commercial status of ISV.

It is worth noting that, while DOE's response to the testing incidents was one of significant withdrawal of support, EPA, in approximately the same time frame, recognized the need to accept and expect some reasonable difficulties with innovative technology testing if their efforts to implement innovative technologies were to be successful. EPA issued a policy to this effect that not only recognized the risk of some failures as important lessons learned, but actually encouraging the selection of innovative technologies despite such risks.

Lessons Learned

Geosafe has encountered and overcome a significant number of barriers in the process of commercializing the ISV technology. The Company has now operated the technology on a commercial basis for more than two years. The future of the technology is very promising. With "perfect 20:20 hindsight", Geosafe suggests the following principles should be considered strongly by parties commercializing innovative technologies:

- Establish a conservative applications envelope for commercial applications of the technology; and do not take projects that involve opening the envelope further without first ensuring all parties understand the experimental and uncertain nature of the outcome.

- Do not risk your company's ability to perform profitable commercial operations on experimental, envelope-widening projects that could damage that capability.

- Make great effort not to let various difficulties destroy the company's marketplace momentum.

- Do not make major commitments to personnel staffing and large-scale field equipment until firm contracts are in hand.

- Develop marketplace and regulatory awareness as broadly as possible, taking care not to oversell the capabilities of the technology. This requires some means of periodic information distribution, and means to keep the awareness current as the technology status changes.

- Recognize that detractors will probably attempt to block your marketplace efforts, and that resources will have to be spent countering their efforts. A strong information distribution program, including publication of performance results, is a good defense against detractors.

- Take great care, even at extra expense, to negotiate contracts that are appropriate for the degree of uncertainty involved in a project. It may be better not to have a contract than to have one that is one-sided, particularly where there exist uncertainties regarding the technical aspects of the project. Ensure that cost-reimbursable standby provisions are included against the possibility of client- or regulatory-imposed delays.

- Recognize that the guidance followed by Government contracting officers will be stringently followed regardless of the spirit of compromise or support indicated by non-contracts type personnel. If tolerance of unforeseen difficulties is to be accepted, such should be defined in the contract.

- Be financially prepared for commercialization time and cost estimates to stretch out beyond seemingly reasonable estimates. This requires experienced investors with a long-term outlook and capability.

ACKNOWLEDGEMENTS

Geosafe acknowledges the importance of the DOE technology development program, and the strong initial support directed to the transfer of the ISV technology for commercial application, as well as ongoing support from the local Environmental Restoration program at Oak Ridge. Geosafe also notes the good support of ISV and other innovative technologies provided by many organizations within EPA and various state regulatory organizations (e.g., Michigan, Washington, Utah). EPA's Technology Innovation Office, SITE Demonstration Program, and Risk Reduction Engineering Laboratory, have been most helpful in supporting the introduction of ISV. EPA was also instrumental in developing independently qualified data on the ISV process that put to rest the unsubstantiated competitor claims mentioned earlier. In addition, the industry media, with the exception of the few who have an agenda beyond providing public information, have been helpful in developing marketplace and regulatory community awareness of the technology. Geosafe is also appreciative of the various "ISV champions" that have emerged within various industry organizations; for without these champions, few innovative technologies would attain commercial success.

THE NEED FOR ADVANCED CHARACTERIZATION TECHNIQUES IN PRODUCT MANUFACTURING: A CASE STUDY ON CERAMIC MATRIX COMPOSITES

Linda M. Braun, National Institute of Standards and Technology, Gaithersburg, MD 20899

ABSTRACT

The capability of industry to manufacture reliable and cost-effective fiber-reinforced ceramic matrix composites requires detailed characterization and control of the composite properties. Advanced characterization techniques, including *in situ* characterization of material properties, offer the potential for process control, quality control, and lifetime monitoring. Advanced characterization techniques currently being developed to aid in the manufacture and scientific understanding of ceramic components are discussed.

INTRODUCTION

The need for materials that are lighter, stronger, more corrosion resistant, and capable of performing at elevated temperatures has led to the development of fiber-reinforced ceramic matrix composites (CMC's). In these complex materials, the fiber/matrix interface controls two key properties of interest, i.e. toughness and damage tolerance[1,2]. Degradation of the fiber/matrix interface during use, at elevated temperatures, has been shown to be a serious problem. Due to their complex microstructure, detection of critical strength controlling flaws and damage in CMC's is very difficult. In addition, the capability of industry to manufacture reliable and cost-effective CMC components requires control of the physical properties (mechanical and thermal) of the final component. To this end, advanced characterization techniques, including *in situ* characterization, have the potential to aid ceramic materials and component manufacturers in achieving the goal of cost-effective, reliable ceramic parts. Characterization techniques designed to monitor the all important fiber/matrix interface during fabrication and during application of the component, would greatly benefit the composites industry. Techniques designed to detect flaws and damage would also be of great benefit. Research scientists are developing advanced characterization techniques to aid in the manufacture of ceramic components. This paper will give a brief discussion of three specific examples of characterization techniques that are currently under development. The examples are specific to fiber-reinforced ceramic matrix composite manufacturing, but if successful, will be useful for the fabrication of a variety of ceramic components.

ADVANCED CHARACTERIZATION TECHNIQUES

As previously mentioned, manufacture of reliable and cost-effective CMC components requires control of the physical properties of the final component. Some physical properties commonly measured by ceramic composite manufacturers include: density or weight gain during nitridation, phase content, flexure, tensile, and burst strengths, thermal properties, physical dimensions, and visual inspection. The challenges for incorporation of advanced characterization methodologies are: 1) maintenance of component cost, 2) implementation of a technique that the customer trusts, 3) implementation of a technique that is easy to perform, and 4) suitability of techniques to heterogeneous large components. The goal is to meet these challenges, and to provide real-time monitoring of the fiber/matrix interface with respect to stresses, impurities, delaminations, and porosity.

Advanced characterization techniques, to aid in ceramic component manufacturing, can be divided into two categories. The first category can be termed "Intelligent Processing", and the second "Non-Destructive Characterization." The first category, intelligent processing, holds the most promise for introduction of *in situ* characterization tools for process control. Non-destructive characterization on the other hand, can aid in quality control and lifetime monitoring of the finished component. Three examples of *in situ* monitoring and non-destructive characterization which could have direct application to the fiber composite community are discussed below.

Intelligent Processing

In situ sensors offer the potential for on-line characterization and process control. A technique which uses a reinforcing fiber, inherent in unidirectionally fiber reinforced composites, to sense characteristics of the fiber/matrix interface is currently under development[3]. In this technique, an ultrasonic wave is imparted to the fiber, and propagates along the fiber. When the fiber is unconstrained (in free space), the ultrasonic wave travels unimpeded through the fiber. When the fiber is constrained within a matrix material, some of the ultrasonic energy leaks across the fiber-matrix interface. The resulting ultrasonic signal which propagates through the fiber is thereby influenced, in its velocity and frequency content, by the nature of the interface. Ultrasonic waves can be launched and detected via contact piezoelectric transducers or detected optically with a laser based interferometric system. This second approach allows the "sensor" to be used at elevated temperatures and pressures such as those used during composite processing[3]. This technique provides information about residual stresses and binder burnout during the processing of fiber-reinforced composite materials. Impurities segregating at interfaces will affect electrical, thermal, and mechanical (bonding) properties. The ability to monitor the fiber/matrix interface *in situ* will aid in the scientific understanding of the critical role of the interface in controlling the properties of the composite.

Another advanced characterization technique is micro-Raman spectroscopy. This technique has been used extensively as a powerful analytical and molecular spectroscopic technique for the identification of contaminants. Raman spectroscopy has been used to identify residual silicon in fiber-reinforced ceramic composites. Recently, this technique has been used to characterize semiconductors.[4] The high spatial resolution of micro-Raman spectroscopy makes it an ideal candidate for *in situ* characterization and process control of a variety of components, i.e. structural, electronic, and photonic. In fact, Raman scattering has been used successfully for real-time monitoring and control of thin-film growth during processing in various environments.[5] This same technology should be transferable to real-time monitoring of a fiber/matrix interface in ceramic composites. It should be noted that a transparent matrix is a requirement in order to use this technique.

Non-Destructive Characterization

Successful implementation of ceramic matrix composites in advanced engineering designs will greatly benefit from the development of non-destructive characterization methods. These methods will be used for quality control as well as to provide data such that, coupled to life-time prediction models, reliable decisions on component replacement can be made. Many application environments of composite systems subject the materials to conditions which may cause failure from thermal shock, delaminations, or oxidation of the critical fiber/matrix interface.

Methods to inspect non-destructively or characterize the materials such that lifetimes can be predicted, are needed before the materials can be utilized to their full potential.

Non-destructive characterization techniques which utilize infrared imaging technology to detect and characterize the "flaws" of interest, are currently being developed.[6] The goal is to obtain full two-dimensional thermal diffusivity maps of ceramic fiber-reinforced composite components. The thermal diffusivity images can be compared both before and after component (e.g. engine combustion liner) operation. Delaminations and porosity can be detected using this technique. Imaging technologies offer the potential for critical "flaws" and damage to be detected prior to total component failure. Another advantage of imaging techniques is that both the manufacturer and the customer can see discrete variations, and therefore, trust the results.

SUMMARY

Advanced characterization techniques are currently being developed that will aid in the manufacture of cost-effective and reliable ceramic components. It should be noted, that the techniques described above are research tools at present. However, advanced characterization techniques offer the potential for 1) improved, more efficient processing and process control, 2) quality control during fabrication and of the final component, and 3) lifetime monitoring of the final component. *In situ* characterization techniques will allow for process control during fabrication, and hopefully aid in reducing the cost of ceramic components.

ACKNOWLEDGEMENTS

The author wishes to thank Dr. Renee M. Kent of Wright Patterson Air Force Base and Dr. Bill Ellingson of Argonne National Laboratory for useful discussions of their work. The DOE Office of Industrial Technology is gratefully acknowledged for support under contract #DE-AI05-92OR22014.

REFERENCES

1 H.C. Cao, E. Bischoff, O. Sbaizero, M. Ruhle, A.G. Evans, D.B. Marshall, and J.J. Brennan, "Effect of Interfaces on the Properties of Fiber-Reinforced Ceramics," J. Am. Ceram. Soc., 73[6]1691-99 (1990).

2 R.F. Cooper and K. Chyung, "Structure and Chemistry of Fibre-Matrix

Interfaces in SiC Fibre-Reinforced Glass-Ceramic Composites: An Electron Microscopy Study," J. Mater. Sci., 22, 3148-60 (1987).
[3] R.M. Kent, Wright Patterson Air Force Base, Dayton, OH private communication.
[4] S. Perkowitz, "Optical Characterization of Semiconductors: Infrared, Raman, and Photoluminescence Spectroscopy," Academic Press: San Diego, 1993, Chapter 6.
[5] V.A. Yakovlev, "How to Characterize Thin Films for Real-Time Process Control," Laser Focus World, pp 83-89, August 1993.
[6] W.A. Ellingson, N. Thiemann, and K. Durkin, "Suitable, Industry-required, Nondestructive Characterization Methodologies for CFCCs," DOE Office of Industrial Technology, CFCC Technical Highlight, 10/05/93.

KEYWORD AND AUTHOR INDEX